Polluted Politics

This book describes the politically charged afterlife of Israeli electronics gathered by and processed in a cluster of rural Palestinian villages that has emerged as an informal regional e-waste hub. As with many such hubs throughout the Global South, rudimentary recycling practices represent a remarkable entrepreneurial means of livelihood amid poverty and constraint, which generates staggering damage to local health and the environment, with tensions between these reaching a breaking point. John-Michael Davis and Yaakov Garb draw on a decade of community-based action research with and within these villages to contextualize the emergence, realities, and future options of the Palestinian hub within both the geopolitical realities of Israel's occupation of the West Bank and shifting understandings of e-waste and recycling dynamics and policies globally. Their stories and analysis are a poignant window into this troubled region and a key sustainability challenge in a polarized globalized world.

JOHN-MICHAEL DAVIS is an assistant professor in The Global School at the Worcester Polytechnic Institute, Massachusetts.

YAAKOV GARB is a professor of interdisciplinary environmental studies at Ben-Gurion University of the Negev.

The Global Middle East

The Global Middle East series seeks to broaden and deconstruct the geographical boundaries of the "Middle East" as a concept to include North Africa, Central and South Asia, and diaspora communities in Western Europe and North America. The series features fresh scholarship that employs theoretically rigorous and innovative methodological frameworks resonating across relevant disciplines in the humanities and the social sciences. In particular, the general editors welcome approaches that focus on mobility, the erosion of nation-state structures, travelling ideas and theories, transcendental techno-politics, the decentralization of grand narratives, and the dislocation of ideologies inspired by popular movements. The series will also consider translations of works by authors in these regions whose ideas are salient to global scholarly trends but have yet to be introduced to the Anglophone academy.

Other books in the series:

1. *Transnationalism in Iranian Political Thought: The Life and Times of Ahmad Fardid*, Ali Mirsepassi
2. *Psycho-nationalism: Global Thought, Iranian Imaginations*, Arshin Adib-Moghaddam
3. *Iranian Cosmopolitanism: A Cinematic History*, Golbarg Rekabtalaei
4. *Money, Markets and Monarchies: The Gulf Cooperation Council and the Political Economy of the Contemporary Middle East*, Adam Hanieh

5. *Iran's Troubled Modernity: Debating Ahmad Fardid's Legacy*, Ali Mirsepassi
6. *Foreign Policy as Nation Making: Turkey and Egypt in the Cold War*, Reem Abou-El-Fadl
7. *Revolution and Its Discontents: Political Thought and Reform in Iran*, Eskandar Sadeghi-Boroujerdi
8. *Creating the Modern Iranian Woman: Popular Culture between Two Revolutions*, Liora Hendelman-Baavur
9. *Iran's Quiet Revolution: The Downfall of the Pahlavi State*, Ali Mirsepassi
10. *Reversing the Colonial Gaze: Persian Travelers Abroad*, Hamid Dabashi
11. *Israel's Jewish Identity Crisis: State and Politics in the Middle East*, Yaacov Yadgar
12. *Temporary Marriage in Iran: Gender and Body Politics in Modern Persian Film and Literature*, Claudia Yaghoobi
13. *Cosmopolitan Radicalism: The Visual Politics of Beirut's Global Sixties*, Zeina Maasri
14. *Anticolonial Afterlives in Egypt: The Politics of Hegemony*, Sara Salem
15. *What Is Iran? Domestic Politics and International Relations*, Arshin Adib-Moghaddam
16. *Art and the Arab Spring: Aesthetics of Revolution and Resistance in Tunisia and Beyond*, Siobhán Shilton
17. *Tunisia's Modern Woman: Nation-Building and State Feminism in the Global 1960s*, Amy Aisen Kallander
18. *Global 1979: Geographies and Histories of the Iranian Revolution*, Arang Keshavarzian and Ali Mirsepassi
19. *Fixing Stories: Local Newsmaking and International Media in Turkey and Syria*, Noah Amir Arjomand
20. *Schooling the Nation: Education and Everyday Politics in Egypt*, Hania Sobhy
21. *Violence and Representation in the Arab Uprisings*, Benoît Challand
22. *A Social History of Modern Tehran: Space, Power, and the City*, Ashkan Rezvani Naraghi

23. *An Iranian Childhood: Rethinking History and Memory*, Hamid Dabashi
24. *Heroes to Hostages: America and Iran, 1800–1988*, Firoozeh Kashani-Sabet
25. *The Making of Persianate Modernity: Language and Literary History between Iran and India*, Alexander Jabbari
26. *Mandatory Madness: Colonial Psychiatry and Mental Illness in British Mandate Palestine*, Chris Sandal-Wilson

Polluted Politics

The Development of an Israeli-Palestinian E-Waste Economy

JOHN-MICHAEL DAVIS
Worcester Polytechnic Institute

YAAKOV GARB
Ben-Gurion University of the Negev

CAMBRIDGE
UNIVERSITY PRESS

CAMBRIDGE
UNIVERSITY PRESS

Shaftesbury Road, Cambridge CB2 8EA, United Kingdom

One Liberty Plaza, 20th Floor, New York, NY 10006, USA

477 Williamstown Road, Port Melbourne, VIC 3207, Australia

314–321, 3rd Floor, Plot 3, Splendor Forum, Jasola District Centre, New Delhi – 110025, India

103 Penang Road, #05-06/07, Visioncrest Commercial, Singapore 238467

Cambridge University Press is part of Cambridge University Press & Assessment, a department of the University of Cambridge.

We share the University's mission to contribute to society through the pursuit of education, learning and research at the highest international levels of excellence.

www.cambridge.org
Information on this title: www.cambridge.org/9781009483636

DOI: 10.1017/9781009483629

When citing this work, please include a reference to the DOI 10.1017/9781009483629

First published 2024

A catalogue record for this publication is available from the British Library.

A Cataloging-in-Publication data record for this book is available from the Library of Congress

ISBN 978-1-009-48363-6 Hardback

Contents

Figures

Preface: Black Rain: An Entry Point

This book is about the politically charged afterlife of Israeli electronics in a Palestinian electrical and electronic waste (e-waste) hub. The story takes place in the West Bank, where a substantial e-waste and used materials processing hub emerged in the early 2000s, centralized in a string of rural Palestinian villages known collectively as the "West Line" in the rolling hills of southwest Hebron (Beit Awwa, Deir Samet, Al Kum, and Idhna). On the morning of most workdays since then, a fleet of nearly 100 trucks set out from these Palestinian villages and cross over into Israel. Toward sunset, the valuable scrap metal and e-waste they manage to amass there is smuggled across the border, as the key input for a multifaceted recycling network of refurbishment and dismantling operations. In a setting with limited options for livelihoods remaining after prolonged Israeli occupation of the West Bank, this cross-border trade has sustained an entire local economy by processing Israeli rubbish. At the same time, rudimentary recycling methods have transformed this pastoral countryside into one of the most severely polluted regions throughout Israel and Palestine. The jarring juxtaposition of this landscape reflects the tensions we discuss in this book. Scenes of rural poverty are spotted alongside impressive villas owned by those who have managed to tap into this unexpected source of value; agricultural fields that were once the economic backbone of these villages are now scarred with blackened burn sites, piles of discarded scrap, and a sky frequently covered with billowing black smoke.

Remarkably, the scope and severity of this cross-border e-waste economy remained ostensibly invisible – evading sustained attention from Israeli and Palestinian national governments as well as nongovernmental organizations (NGOs) and media outlets for well over a decade – and might have remained so for many years more were it not for a series of serendipitous events.[1] In the summer of 2011, one of us

[1] A note on terminology. Electronic and electrical waste is commonly referred to as e-waste, waste electrical and electronic equipment (WEEE), and discarded

(Yaakov Garb) was working with a McGill University intern, Sarah Mendelsohn, to understand the use of household water technologies and sources in the Hebron Governorate. In investigating household rainwater collection patterns using Palestinian national statistics, Garb noted that the West Line villages appeared as an outlier, relying much less on rooftop rainwater collection than neighboring villages with similar precipitation levels. Follow-up interviews conducted by Mendelsohn and Garb in the West Line villages seemed to point to a disturbing cause – residents refrained from rainwater collection due to the "black rain" caused by excessive e-waste burning. While black rain did not literally fall from the sky, the soot from excessive e-waste burning settled on rooftops over the course of long dry summers and would eventually wash into collection cisterns during the fall rainy season and contaminate water supplies – hence the term "black rain." Beyond this, villagers spoke of the choking smoke, which fouled laundry and sometimes led people to sleep with closed windows and wet towels under the doors to reduce the nuisance. Garb and Mendelsohn wrote a proposal and received a small grant from the Belgian Consulate to explore this phenomenon, and began interviewing a range of stakeholders in the e-waste economy and those affected by it. It soon became clear that this was a serious and long-standing problem that deserved more sustained study and interventions.

The following summer (2012), Garb was joined by John-Michael Davis on a similar summer internship, and this was devoted specifically to expanding this research. The Green Land Society for Health Development (a Palestinian health NGO based in the city of Hebron) provided an Arabic–English translator from Dura, a few kilometers east of the West Line villages. For several weeks, Davis resided in Dura and spent his days walking and hitchhiking around the West Line

electronic and electrical devices (DEED) in the academic literature. E-waste and WEEE are typically used interchangeably, whereas DEED is often used to indicate that used electronics are not "waste" but still have value. Throughout this book, we use the descriptive term "e-waste," which is more commonly used in both the literature and policies and in the Israel–Palestine context. It is also worth mentioning that e-waste hubs, such as the West Line villages, typically process a mixture of e-waste and used materials. E-waste is the most significant (by quantity and value) second-hand resource imported to the West Line villages, but is not the only resource as other used materials such as furniture, scrap metal, and used automobiles comprise a significant, and often overlooked, segment of the industry.

villages to flesh out more details of the dynamics and forces that shape this e-waste industry and the associated economic and environmental consequences. His time in the villages also surfaced the local cultural, socioeconomic, and political undercurrents that polarized certain social groups when discussing the e-waste economy. These efforts soon shifted their geographic and institutional base to the West Line e-waste hub itself, where Garb and Davis increasingly worked directly with the local inhabitants involved in the scrap sector or exposed to its consequences.

Our initial field research and understanding of the West Line e-waste economy were supplemented and reinforced by external information sources that allowed us to assess the scale and implications of the phenomenon: a comprehensive mapping of e-waste burn sites (2012–13), a prior study of regional freight movement (2007–8), and the growing literature on the environmental and health impacts of informal e-waste processing (2011–13). Since burn sites were a key expression of e-waste processing and were readily visible from the air, Garb assembled a time series of high-resolution aerial photographs of the area from 1999 to 2011 and devised a system with Davis who manually identified and field-validated burn sites throughout the Hebron Governorate in each year. This mapping produced a systematic spatial assessment of the rise, clustered location, and current extent of this phenomenon (Davis & Garb, 2018a). Research from some years earlier showed the persistence and stability of this sector. In 2007–2008, Garb had conducted a large systematic cordon survey and flow analysis of freight transport to and from the entire Hebron Governorate on behalf of a coalition of Palestinian and Israeli stakeholders (Garb, 2008). A reanalysis of these data focused specifically on the vehicles bringing scrap to and from the West Line villages showed that the patterns and volumes of scrap we were seeing had been in place nearly a decade earlier, in the late 1990s. Finally, we began reading the scientific literature on the environmental and health consequences of e-waste burning, which had grown from a couple of dozen articles annually when these flows had started in our area, to almost five times that when we began examining them in earnest, and has further quadrupled since then. These data sources, combined with our own active research, began to reveal e-waste to be a sustained cross-border economy between Israel and the West Bank – in fact, one of the largest Palestinian exports to Israel – causing severe health and environmental damages.

As the scale and seriousness of this topic became clearer, we began the community-based research and advocacy efforts described in this book, working primarily with local Palestinian partners and then broader circles of stakeholders to understand and change the situation. As is often the case in community-based research, though seldom discussed (McArdle, 2008), our discovery of and engagement with the West Line villages benefited from contingent and converging events and circumstances. Garb's detection of an anomaly in the spatial distribution of the presence/absence of rainwater collection coincided with having serendipitous personal contacts with someone from one of the villages that appeared as an outlier in the data – a village with very high use of tanker water rather than rainwater collection compared to the neighboring villages – which made it the place to start with in-depth interviewing. This happened to be in the heart of the area of burning, hence the initially incredible accounts we heard of cable burning so voluminous as to lead to "black rain." One can discount such stray rumors, but we were lucky enough to take them seriously, so pulling at that small thread soon led to the whole tangled story of the extensive waste economy and its consequences. The arrival of Davis in Israel the following year (2012) offered another happy conjuncture. As an intern with adequate funding for fieldwork, he was able to work on an open-ended exploratory project, while his conspicuous status as an "international" allowed him exceptional access to rural Palestinian areas and the ability to move safely in areas many consider dangerous. At the same time, as a Hebrew-speaking academic at an Israeli university, Garb functioned as an "insider" to Israeli society, which eased potential difficulties in border crossing and access to Israeli information and networks. Despite the troubled geopolitical context of our work, both of us seemed sufficiently well-intentioned, and preoccupied with apparently esoteric research inquiries to somewhat defuse suspicions of us having malign motivations. As the West Line community gradually accepted our efforts, we were able to navigate between social worlds ranging from army officers who supervised checkpoints and patrols to the waste smugglers whose livelihood depended on evading them.

Despite our access to and acceptance within the West Line community, our inability to fluently speak the native language coupled with our obvious status as outsiders to a close-knit, clan-based community undoubtedly hindered our understanding of the lived realities, perceptions, and development desires of the population we studied. Over time, closer working relationships with local translators, research assistants, and research partners, along with long-term immersion in the West Line villages, helped close these gaps. The importance of these efforts became increasingly clear the more time we spent in the villages as our initial simplistic perceptions and understandings of the history, causes, tensions, and suitable solutions to reform the local e-waste industry evolved and matured in ways we would not have anticipated.

Davis' 2012 internship extended into a master's thesis at Ben Gurion University, supervised by Garb, and then a doctoral thesis from 2013 to 2017 at Memorial University of Newfoundland, co-supervised by Garb and another experienced researcher on e-waste at the global scale, Professor Josh Lepawsky. This long-term engagement on the topic enabled a deeper reflection of the multiple cleavages within the West Line villages and their divergent development aspirations using a growing and more systematic empirical basis. At this point, we had conducted nearly 100 interviews with individuals working in the e-waste industry and the local community along with people on both the input and output sides of the e-waste economy. We had also attended several municipal and civic society meetings on the topic, which made apparent the urgent need and desire from the community to address e-waste contamination. We developed partnerships with West Line residents and civic organizations to collaboratively initiate, design, and facilitate a community-based activist research program, which produced a widely endorsed development trajectory to reform the West Line e-waste industry and a series of projects to achieve this goal. The Swedish International Development Cooperation Agency (Sida), which had followed our work since 2013, began supporting the emerging community-based interventions for the West Line villages with a series of three generously flexible exploration and piloting grants from 2016 to 2018 that enabled us to develop a major grant proposal for a package of integrated interventions to reduce e-waste burning, remediate toxic burn sites, and formalize the West Line e-waste industry. These efforts paralleled and contributed to the emergence of the West Line e-waste economy into greater visibility for national and international institutions. From a local phenomenon given little attention beyond those directly involved, it started to become an object of broader knowledge, reform efforts, and NGO attention, as well as media coverage and political blame.

This book is our attempt to synthesize and reflect on these key years of our joint work on this topic, spanning the period from Davis joining the project in the summer of 2012 until the summer of 2018. During this time, Davis lived in the West Line villages for twenty months over five separate trips; the most substantial period was eleven months in 2015. His last visit was in 2018, which marked a turning point in our community-based research. The proposed Sida-funded project was suspended due to an accumulation of legal and political sticking points between both municipal and national Palestinian government offices and Israeli and Palestinian environmental authorities, which we unpack in the second half of the book. The final nail in the coffin was when the project was officially vetoed by the Chair of the Palestinian Environmental Quality Authority (EQA). Afterward, Davis returned to postdoctoral work in the United States of America; Garb's e-waste work in the area continues, together with co-researchers and community members. After the suspension of the Sida project, which was geared toward a large formal effort at locally driven change, the emphasis of this work has shifted. The advocacy efforts are now more reactive and on a smaller scale, supporting Palestinian viewpoints and interests in the face of external (and usually adverse) policy interventions stemming from Israeli stakeholders and the Palestinian Authority. The research emphasizes ethnographic and historical inquiry into the West Line, as well as geochemical and risk assessment work to inform and leverage future protection and remediation efforts.

In presenting these efforts, this book tells three intertwined stories, each with its own goals and audiences. One is an account of e-waste hubs and hazardous informal industrial clusters more broadly: where and why these emerge, how they function, their consequences, and the need for pathways to change them. Another, to a lesser extent, examines the e-waste system and our efforts to reform as occurring within and emblematic of the post-Oslo Israeli-Palestinian space, with its particular anguished tangle of spatial and political conflicts. E-waste becomes an informative lens to deepen our understanding of this area's political geography, where it has been, and where it is going and, perhaps, to offer insights into some unexpected degrees of freedom in a system that seems relentlessly polarized and tragic. Finally, this is a methodological story, of a particular stance toward research and advocacy that is closely coupled with community development and change processes, and the actors, knowledges, opportunities, and pitfalls of such work.

Though these stories are intertwined, different audiences can read the book for the emphasis most relevant to them. For those interested in the policies and politics of waste, and e-waste in particular, our sustained and integrated consideration of a Palestinian e-waste hub builds on recent efforts by scholars and practitioners who have explored and sought to improve the conditions of other controversial e-waste hubs in now infamous sites of Guiyu, China (Chi et al., 2011; Schulz & Lora-Wainwright, 2019; Wang et al., 2021); Agbogbloshie, Ghana (Grant & Oteng-Ababio, 2021; Oteng-Ababio et al., 2020); Lagos, Nigeria (Nnorom & Osibanjo, 2008a; Osibanjo & Nnorom, 2007; Sullivan, 2014); Bangalore, India (Borthakur & Govind, 2017; Reddy, 2013, 2015); Cape Town, South Africa (Grant, 2019; Lawhon 2012a, 2012b), among others. While the Israeli-Palestinian case study and our policy package to reform the West Line e-waste hub are inflected by local geopolitical, socioeconomic, and cultural circumstances in unique ways, as is the case for any such hub, the Israeli-Palestinian instance offers a remarkable window into dynamics and predicaments broadly relevant for the Global North–Global South e-waste trade and e-waste hubs more generally. For those for whom the Israeli-Palestinian context is central, on the other hand, we hope that the e-waste predicament offers a new and productive lens into the situation and prospects of this geopolitical and economic terrain. Finally, this is a book about community development and change that may be helpful for others engaged in forging a bottom-up redirection of harmful practices and production networks. Such readers might benefit from our account of the evolution of our approach, what worked and what did not, and from our reflections on how these strategies and pilot projects might be scaled up and out – to other places and industries.

The book is based on our extensive findings on the particular e-waste hub we worked with for many years, which is seen as illustrating broader themes: global e-waste dynamics, informal industrial clusters, and the tensions between livelihood and contamination. Our work has shown that, despite its important idiosyncrasies, the Palestinian hub and its relation to Israel have great relevance for other hubs in the Global South (i.e., confined regions where e-waste processing is the primary industry) and their relation to the e-waste exports and policies of the Global North. In fact, despite – and in some respects, because of – its uniqueness, the Palestinian context allowed us a valuable window into the more general dynamics and opportunities to understand and reform Global North–Global South e-waste trade and e-waste hubs. Indeed, the structural logic and set of dilemmas we portray are relevant not only to e-waste hubs but to other kinds of toxic livelihoods within informal industrial clusters. Tanneries, artisanal mining, brick kilns, and charcoal plants are examples of other industrial clusters in the Global South that present a similar dilemma: providing a much-needed livelihood for marginalized and vulnerable populations while causing severe and enduring environmental and public health harm.

There is not enough nuanced attention among the media, NGOs, and policymakers given to this dilemma of toxic livelihoods, which too easily slides into polarizing and caricatured stances. Workers in these industries typically understand but often downplay their exposure to contaminants. The focus for many is survival, earning enough money to feed their families, or for some kind of lucrative income that would be impossible through less hazardous work. Some neighboring residents whose income is derived from other sources tend to villainize those creating environmental hazards. Others are in a more conflicted situation, both resenting and benefiting from the industry. The clearly oppositional viewpoint is more readily resonant with external actors, who often simplify toxic livelihoods, focusing exclusively on environmental consequences with little consideration of the economic importance or the political and historical circumstances that created them in the first place. This flattening undermines understanding, and the forging of workable trajectories for change based on a deeper structural sense of why things are stuck. Motifs of villainy and ignorance only get you so far, and ascribing behaviors to these tend to gloss over important clues to the dynamics and structures at play.

The need to engage with the real tensions of toxic livelihoods experienced by various actors affected by the e-waste industry emerged as a major theme in our work. Our initial research in the West Line villages quickly surfaced these and the potential harm that can result from conducting research and advocating change for, rather than with, the community. We witnessed well-intentioned outside organizations spend a couple of days seeing smoke and toxic ash, speaking to the small minority of locals organized along familiar thematic and institutional lines in a way that elevated one side of the equation. Finding something that looks like a local environmental NGO is easier than interviewing scrap smugglers or going out in the field with waste burners. Recognizing that we bring our own implicit biases and assumptions concerning ways to improve the conditions of the West Line villages, we resisted the urge to suggest solutions we thought to be most appropriate, and relied on a variant participatory action research (PAR) approach that was intended to more fully encompass community breadth and concerns. Participatory action research is an approach to inquiry that emphasizes engagement and question-posing, working together with participative communities (often emphasizing at-risk stakeholders) to focus on issues they find significant and pressing (Brydon-Miller et al., 2003). Inquiry based on PAR principles understands the world through collective efforts to transform it, collaboratively and reflectively. The process typically engages communities in systematic cycles of action and reflection. In action phases, communities test practices and gather evidence; in reflection phases, they make sense of their experience together and plan further actions. By integrating knowing and acting, these alternating phases enable PAR to bridge the gap between distanced understanding and engaged change-making that can limit more traditional social research.

It is important to emphasize, however, that while our research questions and approach were participatory and community-based, it would be wrong to call the process inherently "community-driven," since much of the initiative and direction came from us, as outsiders. Indeed, this is widely recognized in the literature as a reality of PAR as, in practice, control of the research is rarely delegated to the community, nor would they necessarily want it (Cornwall & Jewkes, 1995). Researchers typically introduce and initiate PAR as a means to promote social change within communities, since it is rare for a community to be organized together, seek out a form of PAR to solve pressing issues in their community, and solicit a researcher to facilitate this process (Maguire, 1987). Paradoxically, most PAR projects would likely not occur without the initiative of external researchers who have the time, skill, resources, and commitment to drive them forward (Reason, 1994). We will discuss this paradox in Chapter 4, where we suggest that the forceful involvement of outsiders can sometimes navigate the pitfalls not only of externally imposed policies, but of apparently endogenous ones. Too often, the latter can rapidly be captured by local or para-local elites, functioning as development brokers, who claim to represent local interests as they produce a kind of donor-friendly pseudo-participation that converges with top-down global environmental policies.

Our PAR approach in the West Line evolved over time, balancing research and action at different intensities. We iteratively progressed from simply gaining an understanding of the scope, dynamics, and impacts of the e-waste industry (2011–14) to strategically envisioning community desired futures (2015), which led to a sustained action phase (2016–18) that developed and implemented community-defined solutions that have since shifted to continued, albeit less active, advocacy efforts (see Figure P.1). Our community engagement in the West Line comprised immersion in the villages, interviewing hundreds of residents, and partnering with eight diverse "community representatives." Together, we co-articulated a community development vision, which then served as the basis when meeting with relevant stakeholders outside the villages who influence the e-waste economy (e.g., government agencies, donors, NGOs, lawyers, academics, formal e-waste recyclers, etc.). This broad and sustained engagement helped us understand the varied perspectives of the West Line e-waste industry and the challenges of the wicked problems the village communities face.

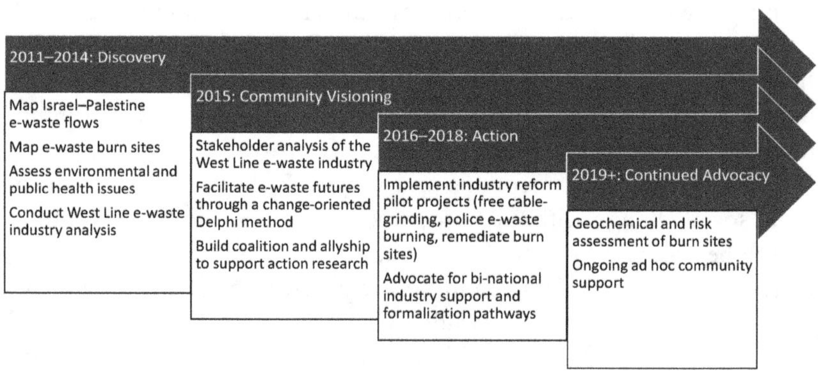

Figure P.1 A sequential graphic depicting the trajectory of our community-based research in the West Line villages. The overlapping and continuation of each phase symbolize our iterative approach and how we engaged in all research phases simultaneously, but at different levels of focus and intensity.

As we share throughout this book, we learned about the e-waste burning niche, for example, by "job shadowing" Muhammad,[2] who earns his main income by purchasing cables extracted from e-waste items, burning them, and then selling the remaining copper. We learned from Layla, the head of the local women's society, how women in the villages work to minimize exposure to e-waste contaminants in their houses. From farmers we learned about the many ways that e-waste burning has damaged local agriculture and livestock. We learned about the history, evolution, and barriers to reforming the e-waste industry from Ali, who has owned an e-waste dismantling scrapyard for nearly two decades. We learned about the near-insurmountable barriers to establishing a formal cable recycling business model from Safa Recycling, the lone formal operating e-waste recycling facility in the West Line villages. And through extensive discussions with the Israeli Civil Administration and the Palestinian EQA, we learned firsthand how the current geopolitical situation has created the conditions for an illicit cross-border e-waste economy on the one hand, and how it impedes productive responses to joint environmental issues on the other.

The participatory and local grounding of both our research and advocacy efforts was central to our engagement in the West Line villages and to this book. As we discuss in detail in the following chapters, informal and bottom-up entrepreneurial centers emerge to offer a precious form of livelihood and value extraction in a landscape of considerable constraint but do so in a way that also causes real damage at local and even global scales. Yet, at the same time, commonplace global and top-down framings of and responses to these hubs can further damage and disempower these communities. Indeed, though these hubs embody the same remarkable degrees of vertical integration, resilience, ingenuity, and adaptiveness as formal clusters, they are not esteemed as vital industrial clusters, but denigrated as *ad hoc* sites of pirate activities by policymakers and media outlets. This stance is embodied in media portrayals of "digital dumpsites" whose backwardness casts them as passive or victimized destinations for the ruthless dumping of more developed countries. While formal industrial clusters are celebrated and cultivated, informal clusters are misread

[2] This name and other names used throughout this book are pseudonyms used to protect identities.

and stigmatized, leading to policy responses that usually range between ineffectual at best and belligerently destructive at worst (Davis & Garb, 2015).

The participatory approach we used held special relevance for the domain of e-waste, where the major regulatory frameworks were shaped and implemented in a top-down fashion. The now-canonical Extended Producer Responsibility (EPR)-based national e-waste management, the regulation of international e-waste flows through the Basel Convention, and the more stringent Ban Amendment in particular were all crafted in the name of environmental improvement of e-waste hubs, but with little input from these places and communities (Kunz et al., 2014, 2018; Puckett et al., 2002, 2005). We suggest that this may be part of the reason current e-waste policies all too often remain as rhetorical declarations, which do little to change the status quo at the local level. In fact, as we show in Chapter 3, such top-down e-waste management policies can, in fact, galvanize rent-seeking and the upward reassignment of value in the commodity chain toward stronger formal actors. In this way, they degrade, rather than advance, the social and environmental conditions of most actors at the lower ends of the e-waste economy in these hubs.

By emerging bottom-up, our book joins and advances an emerging body of interdisciplinary literature that has critiqued and offered correctives to such one-dimensional descriptions of and responses to e-waste hubs (Corwin, 2018; Kojima et al., 2009; Lepawsky et al., 2017; Manhart et al., 2011; Osibanjo & Nnorom, 2007; Reddy, 2015; Wang et al., 2012). Within the Israel-Palestine context, we describe a more complex reality than purported narratives of "toxic dumping" or a "pirate" industry (ARIJ, 2012; GLSHD, 2013), demonstrating the entrepreneurial inventiveness of Palestinian recyclers, who managed to draw e-waste to the West Line villages, extract valuable metals, and sell them back to Israeli buyers, creating one of the largest export sectors to Israel of the entire West Bank. Similarly, we suggest that our community-based approach to evolving a development vision of those most affected by and familiar with the promise and peril of existing practices yields policies and interventions likely to be more just and effective. We generalize the experience of our bottom-up approach to suggest that a hub-centered approach offers an integrative perspective to informal e-waste economies that is more likely to bring about deep change, rather than its appearance.

The Palestinian case offered special opportunities, as well as special challenges, to develop and test a comprehensive and integrated package of solutions that sequentially provides economically attractive alternatives to harmful dismantling practices, remediates environmental contamination, and offers pathways and incentives to integrate informal recyclers into formal systems. The striking geographic adjacency of two societies with great economic disparity, and the closed nature of the Israeli-Palestinian e-waste system, made our study logistically and methodologically feasible, and the exceptionally sensitive and patient conceptual and material support we received from Sida allowed us a rare opportunity to develop and test the emerging models with the care and thoroughness they deserve. At the same time, these efforts took place in the context of the continued unraveling of the Oslo Accords between Israel and the Palestinian Liberation Organization (PLO). These Accords marked the start of a peace process and created a Palestinian Authority that would transition from limited self-governance in areas of the West Bank and Gaza toward a negotiated final status agreement. By the time we began working, however, the situation was deeply stuck. An entrenched right-wing Israeli government backed by an increasingly explicit annexationist drift of Israeli politics faced a Palestinian Authority, increasingly regarded as paralyzed and corrupt, caught between the Israeli government and donors to which it owed its existence and the population it was attempting to govern in impossible circumstances. This was not an auspicious moment to attempt reform of a cross-border e-waste economy that owed its existence to this tangle, using measures whose scale-up depended on coordination of two under-resourced and distracted administrations cruelly and asymmetrically pitted against one another. Our story, therefore, is of possibility and movement, but also tragic blockage, making us even more aware of the importance of telling it in a way that allows translation into other contexts.

Acknowledgments

An eight-year community-based action-oriented research project is only possible with the generous support, contributions, and goodwill of many people across social groups. We extend our heartfelt thanks to everyone associated with the Israeli-Palestinian e-waste economy whom we met, interviewed, observed, surveyed, sampled, worked with, and befriended.

We spent years living and researching in the West Line villages, primarily in Beit Awwa. The hospitality we received was like no other. At times, the invitations for meals and tea were so frequent it became challenging to accommodate each offer. It would take a book in its own right to list everyone by name and express our gratitude. We thank those at the core of this industry, the many scrapyard workers and owners who eventually warmed up to us and shared the many facets of their livelihood. There were several West Line residents who were key supporters of this project: Sara Awadi, Saed Sweity, Atef Awadi, Morad Sweity, Sameera Albtran, Morad Sweity, Maysoon Sweity, Bassam Aleswed, and Ismail Abujhaisheh. Many others (especially Sameer, Hamam, A. Amran) generously gave us insights into their worlds, whether dismantling waste and burning cables, or protecting their livelihoods against these. We are particularly grateful for the Sweity family. (While there are literally hundreds of "Sweity families" in Beit Awwa, they know which one we refer to.) The Sweity family provided everything we needed (and more) to conduct our research: a place to live, translation, transportation, companionship, cultural lessons, and everything in between. When we think of this book, we think of them.

We are indebted to the many reliable and knowledgeable translators and research assistants who worked closely with us in the West Bank. We were fortunate to find not only competent translators from the West Line villages, but young people who were inspired to improve the economic and environmental conditions of their hometown. Without their gracious and eager support, our community-based research would be impossible. We are particularly thankful to work with Mohammad Dehaidel, Hussan Sweity, Muhammad Sweity, Abdullah Sweity, Salem Sweity, and Sulyman Sweity. We also benefited from the contributions of several interns from McGill University – Sarah Mendelsohn, Alizé Carrère, Christopher Nzediegwu and Colin LaRouche. Each intern brought their own enthusiasm, experiences, and research interests, which helped us understand various aspects of the Israel-Palestine e-waste economy.

Working with colleagues, collaborators, and mentors on *Polluted Politics* has been essential to deepen our understanding of the research and associated advocacy we conducted. At Memorial University of Newfoundland, we benefited from the diverse collection of graduate students Professor Josh Lepawsky brought together to study e-waste in various regions of the world. It was a fortuitous opportunity to work so closely with Dr. Grace Akese (sharing an office with Davis), who researched informal e-waste economies in Agbogbloshie, Ghana, which generated comparative learning between our respective fieldwork sites. Josh provided steadfast guidance during fieldwork and made invaluable linkages to the discard studies literature, encouraging us to think more broadly about what is the right thing to do with e-waste. We are also thankful for the many colleagues who engaged in thoughtful discussions and generously provided comments on articles/chapters related to the book that advanced our understanding of the complexities of the Israeli-Palestinian e-waste system and e-waste hubs more broadly. We thank Scott Jiusto, Charles Mather, Liam Swiss, Julie Trottier, Nelly LeBlond, and Jesse Ribot.

We received generous financial support that provided us with the time and resources to conduct the long-term, sustained research we describe in this book. We are thankful for The Norman Zavalkoff Family Foundation Travel Fund for Water Resource Management at McGill University, which initially brought us together on this project. We thank the Social Sciences and Humanities Research Council of Canada (SSHRC) who provided base funding for Davis' PhD and postdoctoral research. In addition to grants for a data server from the Swiss Embassy and funds for exploratory burn site mapping on agricultural land from the Food and Agriculture Organization (FAO), we received three substantial grants from Sida, which enabled us to pilot a package of community-defined interventions to improve the

West Line e-waste industry that became the core of this book. Our understanding of e-waste interventions – what works, and what does not, and the political dynamics at play – would not be as clear without the resources to learn from doing, and the exceptional stewardship of Johan and Gunnar. We also thank the Global Lab Fellowship at Worcester Polytechnic Institute, which provided the dedicated time necessary to complete the book along with audiovisual support. More than once, the Fischer family allowed us to meet real-time developments by advancing any critical expenses before assurance that they would be covered by less flexible sources.

Transportation within and between Israel and the West Bank has been a challenge since we began our fieldwork. Eiman, of blessed memory, offered logistic support as the most competent team manager imaginable, and would surely still be central to everything had he not taken that one last late-night drive. Subhi contributed his unparalleled strategic advice and navigation skills. While it was far from perfect, we are grateful for our gray Peugeot 206 that served as our fieldwork car from 2015 to 2017. You were one of the few legally owned and insured cars in the West Line villages and saved us immeasurable time as we were no longer reliant on hitchhiking, shared vans, and walking to conduct fieldwork. Although you were a far cry from reliable, in many ways, that helped us fit in. Whether it was petrol leaking through the back seat, the transmission only working in reverse, or the heat melting away various appendages – your many failures opened opportunities to meet new people with spare car parts and learn firsthand about the generosity of West Line residents. Like us, parts of you will remain in these villages forever.

Finally, and above all else, we thank our families for their enduring love and support.

JM: To Catalina, who lived in Israel and the West Bank with me for a year, your passion and dedication are second to none. Without you, I cannot imagine continuing with this project as long as I have. To Maya and Leo, thank you for being the most welcome distraction and providing a source of excitement and curiosity outside of academia. To Scrappy, possibly the only dog to immigrate from the West Line villages to Canada, may you rest in peace.

YG: This project has long ago moved from being an academic one, binding me to the people and landscapes of the West Line for over a decade. My gratitude begins with the professional, but extends far beyond, to the countless people who entrusted me with their time, knowledge, and homes, infusing me with their own fierce love of the area's landscape, lifeways, and traditions. Know that you deepened my despair at the cursed situation of recent generations here, but also my certainty we will emerge on the other side, together. Closer to home, my parents and siblings, and beloved wife and daughter, who supported my esoteric efforts despite the occasional costs to them. (It is not easy for a respectable teenager to have a father with a tendency to check the labels and innards of discarded washing machines or television sets they pass on the street.) And so many others, whom I hope will forgive or be grateful for my omission of explicit mention.

PART I

Positioning E-Waste Hubs

1 *The Emergence of E-Waste Hubs*

Once one is sensitized to the shadow world of waste, a drive through most areas of the West Bank will reveal ubiquitous dismantled scrap metal piles and the small ashy patches lining roadsides, with telltale remnants of the electrical and electronic components that were burned there. But these traces of small and usually ad hoc pockets of recycling do not prepare one for the landscape-scale imprint of the persistent, pervasive, and massive scale of e-waste and scrap material recycling in the West Line – a string of rural villages located in southwest Hebron along the Israeli border (see Figure 1.2). Here we find an e-waste hub: an organized, durable, and centralized industry that has processed most Israeli e-waste for almost two decades and become a pillar of the local economy.

While the notion of "e-waste hubs" is used casually in the media and academic discourse to refer to scores of similar clusters around the world, this remains a phrase, rather than an analytic category to indicate the special and remarkable social and economic processes in such places. Informal e-waste hubs are analogs of the celebrated "industrial clusters" of the formal economy. Somehow, away from, and sometimes despite, the intentions of governments and formal economic institutions, e-waste hubs have emerged as hosts to thriving integrated concentrations of businesses, managing far-reaching economic and logistic ties through which they process (albeit with methods that cause severe environmental harm) much, if not most, of the world's e-waste. These hubs achieve higher rates of e-waste collection and recycling than some of the most advanced systems with formal recycling policies. According to the most recent assessment of global e-waste quantities and flows from The Global E-waste Monitor, in 2020 only 17.4 percent of the globally generated 53.6 million tonnes of e-waste was "documented to be collected and properly recycled." Of the undocumented 82.6 percent, only 8 percent from higher-income countries goes to residual waste streams, with most of the remaining

74 percent assumed to move through informal channels and destinations, with large environmental and health harms (Forti et al., 2020). Since e-waste is a fairly recent phenomenon, with rapidly shifting composition, sources, and regulatory landscapes, the rapidity of these hubs' emergence and their flexibility in tracking these dynamic economic activities is impressive.

E-waste hubs embody a dark side of global capitalism. A combination of growing global consumption and rapid product obsolescence has made e-waste the fastest growing waste stream worldwide and presents complex disposal challenges due to its internal components combining both valuable metals and toxic substances (StEP, 2016a). E-waste management policies and formal recycling systems have failed to keep pace with increasing e-waste quantities sparking a key and pressing environmental problem, with considerable attention focused on the transboundary movement of discarded electronics from the Global North to e-waste hubs in the Global South. As we discuss in Chapter 3, the prevailing narrative of the e-waste problem frames consumers in the Global North as perpetrators generating excess e-waste that is dumped on marginalized populations in the Global South. Drawing parallels to the outsourcing of manufacturing to overseas destinations where low labor costs and lax regulations allow increased profits (USNEWS, 2014), nongovernmental organization (NGO) and media portrayals of e-waste hubs in Asia and Africa highlight the poor working conditions and excessive pollution in these sites, which are considered to be dumping grounds for the refuse of rich countries (*60 Minutes,* 2008; *Frontline,* 2009; Puckett et al., 2002). This common storyline not only obscures the environmental consequences in the production of electronics and the environmental limits of recycling (Lepawsky, 2018), but also characterizes e-waste hub residents as helpless victims of transboundary environmental injustice. E-waste hubs do not figure as real places nor is the local population seen as political actors with agency; rather, e-waste hubs serve as warning icons to motivate the urgency of instituting formal systems – symbolic markers of distant places and people to be eliminated, saved, or reformed through assimilation within formal systems. This book centers e-waste hubs as a key component of the e-waste dilemma and, through our sustained community-based research in the West Line villages, describes and argues for a hub-centered lens to reframe conceptions of environmental justice within e-waste discourse and policies.

Operationally defining industrial hubs is notoriously difficult, and e-waste hubs are no different. Definitions are often blurry and contentious, but for our purposes, a broad characterization suffices: a confined geographic region where e-waste recycling and related economic activities are the primary industry supported by common markets, technologies, and labor. E-waste hubs operate as informal economies, which contrast with the newly emerging formal economies created through the government-sanctioned marketization of e-waste, primarily through Extended Producer Responsibility (EPR) policies, where e-waste is "owned" and managed by Producer Recycling Organizations (PROs) that require e-waste collection and recycling businesses to be registered and monitored, adhere to e-waste processing standards, and pay taxes. Businesses within e-waste hubs typically adhere to none of these formal arrangements. However, there is a range of intermediary or blended situations, such as businesses that are registered and pay taxes but obtain inputs from undocumented sources on a cash basis or formal recycling companies that rely on informal partners to take and process certain fractions of the waste stream that are marginally profitable within the formal system. While formal e-waste economies are planned and supported by government policies, e-waste hubs have emerged in regions with specific socioeconomic, geopolitical, and cultural characteristics and support diverse economic activities despite government ambivalence or even opposition.

Within the West Bank, small clusters of e-waste processing have emerged in dispersed regions (e.g., Yatta, Qalqilya, Ezaria, Anata, the villages surrounding Nablus, and others[1]), yet only the West Line qualifies as a hub in this sense, on par with others internationally (e.g., Guiyu, China; Agbogbloshie, Ghana; Lagos, Nigeria; Karachi, Pakistan; Moradabad and Seelampur, India) (Doron & Jeffery, 2018; Greenpeace, 2009; Ploumis, 2011. For a review of e-waste hubs, see Auclair et al., 2020. Beyond the impressive quantity of e-waste and used materials processed in the West Line villages and its economic importance both locally and nationally, as described in Chapter 2, this industry is so central to the local economy that villagers consistently estimate that upward of 80 percent of households benefit from it. The

[1] Metals extracted from e-waste (copper, aluminum, and steel) throughout the West Bank are funneled through one of two Arab-Israeli metal traders in East Jerusalem with the capacity to export to global markets (described in Chapter 2). Interviews with accountants in each of these facilities revealed that 75 percent of their inputs come from the West Line villages.

changing daily prices of copper, steel, and aluminum influence the trade in unrelated businesses, such as barbershops and restaurants, and even children can list metal prices and comment on their fluctuations.

We begin our book with an obvious question that will serve as a useful entry point into a more nuanced account of e-waste hubs, both in our case and in others: "why here?" National-level discrepancies between Israeli economies and environmental enforcement regimes prevail throughout the West Bank but do little to explain why this particular rural area emerged as Palestine's "scrap capital." We can ask similar questions at other scales. Globally, developing countries have low worker salaries and lax environmental protection, yet only a few regions in a subset of developing countries have emerged as key destinations for foreign and domestic e-waste. Of course, the emergence of hubs is partly path-dependent, a matter of chance. But beneath the noise, there is a surprising amount of regularity in the characteristics that foster the emergence and clustered growth pattern of these hubs, and these are the subject of this chapter. In what follows, we introduce the factors that facilitated the emergence of the West Line e-waste hub, and how local contexts come to tap into and flourish from this new source of livelihood. We suggest that beneath the particularities of the Israeli-Palestinian case lie broader similarities with other hubs globally.

As we will see, a detailed investigation of the circumstances and agency of e-waste hubs demands pushing back against a familiar and often implicit characterization of these places as dumping grounds reflecting national-scale dynamics within global political-economic forces and flows. This kind of initially intuitive characterization of Global North–to–Global South e-waste flows has, with varying degrees of explicitness, propelled and relied on the pollution haven hypothesis (PHH) (Lepawsky & McNabb, 2010; Pellow, 2007). The PHH posits that pollution-intensive economic activities will obtain a comparative advantage of reduced costs by relocating to jurisdictions with less stringent environmental regulations. In the Israeli-Palestinian case, markedly different labor costs and regulatory vigor are certainly a major part of the story. Yet, there is much that must be added to the PHH narrative.

While the PHH has been mobilized to understand relationships between trade and environmental behavior since the 1970s (Clapp, 2001; Kellenberg, 2010), it has only recently been applied to trade patterns in hazardous waste, including e-waste (Bernard, 2015;

Lepawsky & McNabb, 2010; Lucier & Gureau, 2015; Pellow, 2007). As it gained traction as an overall framework to situate global e-waste trade patterns, this framing informed efforts to prohibit e-waste trade from developed to developing countries, most notably under the purview of the Ban Amendment to the Basel Convention, an international treaty restricting shipments of hazardous waste from developed nations to the developing world.[2]

Despite its policy salience, there have been few empirical attempts to explicate and test the PHH with respect to e-waste flows. In fact, as more studies and empirical materials speaking to e-waste flows and pollution havens emerge, they suggest a greater degree of geographic variability than accounted for by the PHH, demanding a more nuanced portrait of global e-waste flows and their destination sites (Breivik et al., 2014; Efthymiou et al., 2016; Lepawsky & McNabb, 2010). For example, studies mapping global e-waste trade patterns have shown that e-waste movement does not only flow from developed to developing countries, but also includes e-waste trade within and between developing countries (Furniss, 2015; Lepawsky & McNabb, 2010). The patterns, motivations, and agencies mobilizing flows to e-waste hubs extend beyond simple accounts of dumping and flight to jurisdictions with the lowest labor costs and least stringent environmental regulations (Kahhat & Williams, 2010; Lepawsky & Billah, 2011; Reddy, 2015). Thus, the very definition of hubs as disposal sites for global e-waste demands a reconceptualization. These are sites where real value creation and entrepreneurial inventiveness occur as the range of processes extends beyond simple rudimentary extraction to more complex forms of reuse, repair, and material recovery (Grant & Oteng-Ababio, 2013; Lepawsky & Billah, 2011; Lepawsky & Mather, 2011; Reddy, 2016; Schleup et al., 2009; Tong & Wang, 2004). E-waste hubs emerge after years of an accretionary virtuous cycle of growth, which gradually accumulate expertise and tacit knowledge, networks and complex supply chains, specialization, niche diversification and synergies, and auxiliary support systems (which,

[2] Specifically, the BAN Amendment prohibits e-waste trade from Annex VII countries (Organization for Economic Cooperation and Development (OECD), the European Community (EC), and Liechtenstein) to Non-Annex VII countries (see Lepawsky, 2015). At the time of writing this, the Ban Amendment has been accepted by eighty-six countries and the European Union, but has not entered into force, which requires ratification from three-quarters of all member states to the Convention.

conversely, represent sunk costs and spatial lock-in), and are all fed by increasing economies of scale, so that a hub becomes an actor and magnet for e-waste flows. This emergent and even entrepreneurial dynamic is quite different than the dumping ground/pollution haven imagery that guided early treatments of such hubs. Our work adds to this emerging consensus. We argue that the PHH is not so much wrong as far too simplistic – an inadequate universal description of global e-waste flows. The PHH obscures local and contextual factors that, if better understood, would provide more insight and better guidance for e-waste policy. As we describe the dynamics in the West Line e-waste hub below, we also critique and enrich the theoretical conversation around the PHH as the main framework for understanding how and why hubs emerge while laying the groundwork for a different approach to studying them and developing hub-based policies.

As context for our discussion of the predisposing features of the West Line e-waste hub, we begin with a broad overview of the Palestinian economy under Israeli occupation. Specifically, we discuss the circumstances that have undermined Palestinian economic development and inhibited environmental protection, as well as setting the stage for the West Line's repeated sequence of articulation and disarticulation from the Israeli economy that are central to the hub's functioning. We then tease apart the place-based attributes that facilitated this region's engagement with scrap metal and used materials and then its emergence as an e-waste hub. These place-based features are not exceptional to the West Line e-waste hub case, and we illustrate the generalizability of these through brief portraits of how they play out in the two most publicized e-waste hubs globally, Guiyu and Agbogbloshie. Finally, we conclude by showing how this kind of account challenges prevailing narratives of e-waste hubs as pollution havens, or "digital dumpsites," as well as the moral impulses and policy responses this framing favors.

1.1 The Palestinian Economy under Israeli Occupation

Insofar as one can generalize about "developing countries," the West Bank[3] fits within this category but is inflected by Palestine's actively

[3] Because the status and economies of the two areas of contemporary Palestine, Gaza and the West Bank, have become so divergent over the last decade, our discussion is entirely of the latter.

colonial situation, rather than postcolonial, and the occupation is characterized by a spatially contiguous yet strangely perforated and indeterminate border, giving rise to a kind of troubled fusion with unique advantages and woes. The mechanisms of Israel's occupation – extraction of Palestinian resources and labor, and unilateral control of borders and movement – and their contribution to the continued decline of the Palestinian economy are well known (Gordon, 2008; Hever, 2010), so our focus here is on dynamics most relevant for the emergence of the West Line e-waste hub. The structural groundwork for this emergence was laid over the course of three main periods between 1967 and 2004.

1. In the decades following Israel's Occupation in 1967, Palestinians were integrated into the Israeli workforce in agriculture, construction, and other blue-collar jobs, leading to increased blurring of the two economies and flows between them, through a mostly unplanned but steady process. During this time, Palestinian unemployment in the West Bank was below 5 percent (PCBS, 2003).
2. The outbreak of the First Intifada in 1987 interrupted the gradual emergence of this ad hoc integration, while the Oslo Accords of 1993, followed closely by the Protocol on Economic Relations (the Paris Protocol), began to rework the West Bank's spatial and governance structures. The degree to which the apparent turnaround of the Oslo "peace process" masked a continuation and even deepening of Palestinian dependency and arrested development became increasingly clear during the Oslo Accord's stipulated five-year interim period, which ended in 1999 (Naqib, 2003; Roy, 1998).
3. The years of the Oslo process' stasis and dashed aspirations came to a head in 2000 with stalled negotiations and then the outbreak of violent resistance of the Second Intifada. This led to a significant increase in closures and a precipitous halving of the number of registered Palestinian laborers permitted to work in Israel. Over the following years, the aspiration for "separation from the Palestinians" increasingly voiced by Israeli politicians and the public solidified into measures with real bite, undermining the kind of mobility to which a generation of Palestinian workers had grown accustomed. A series of dramatic terror attacks gave a visible rationale for a regime of Israeli work permits, mobility restrictions,

and border closures, which had been taking shape over the Oslo years, while a suspicious Israeli public pressured politicians to accelerate the previously fitful construction of the "separation barrier" between Israel and the West Bank. Work opportunities became increasingly tenuous, which were further undermined by a Palestinian economy that had been crippled by decades of capped opportunities alongside a deep dependence on the Israeli economy, goods, edicts, and labor markets. By 2004, unemployment had risen to be chronically over 20 percent (Khalidi & Taghdisi-Rad, 2009), and poverty reached crisis proportions with 67 percent of Palestinians below the poverty line and 48 percent in extreme poverty (Kawasmi & White, 2010).[4]

Thus, throughout these periods, the Palestinian economy was fused with Israel's economy under uneven terms and then reworked along equally dependent but more strongly separated lines with the establishment of the Palestinian Authority. These dynamics set in place spatial, governance, economic, and social conditions facilitating the emergence of the West Line e-waste hub around 2004. Key among these were the following realities:

1.1.1 Economic Dependency and De-Development

During the decade after the occupation, the Israeli economy absorbed roughly one-third of the Occupied Territories labor force, and the Green Line was more or less porous in economic terms. While Palestinian men served as an "industrial reserve army" (Portugali, 1993), independent Palestinian economic development was stifled, and critical resources such as land, water, and stone/aggregate were controlled and appropriated by Israel. In the wake of the Oslo Accords, the Paris Protocol established the interim-period economic relations between Israel and the Palestinian Authority based on a "customs union" model, which unified some aspects of the two economies (currency, tax relations, etc.), with Israel retaining

[4] The poverty and employment statistics in this paragraph should be taken with caution as the studies cited do not indicate whether they incorporate informal employment into their calculations and neither study provides an operational definition of "poverty" or "unemployment."

unilateral control of key leverage points, such as control of the borders (because of this, Palestinian businesses find it difficult to access international suppliers and markets directly). Palestinian real per capita gross domestic product (GDP) fell 36.1 percent between 1992 and 1996 because of falling aggregate income and rapid population growth. Palestine became an increasingly donor-dependent and service economy: with 38 percent of the Palestinian GDP generated from the service sector, 5 percent generated from agriculture, 15 percent from industry, and foreign aid, which stood at $1.8 billion in 2008, constituting 30 percent of the national GDP (MAS, 2010).

Palestine's economy is substantially dependent on access to permits to enter Israel for employment. The permit regime has functioned as a mode of control over the Palestinian population since it has established and promoted norms of "correct" behavior, which has shaped the daily life and demeanor of Palestinians in line with Israeli interests (Gordon, 2008: 38). Palestinians requesting a permit are sometimes interviewed by an officer of Israel's General Secret Services, also known by its Hebrew acronym as the Shabak or Shin Bet. Access to the requested permit can sometimes be made contingent upon the applicant's willingness to collaborate, which can entail relaying information to Israeli authorities on persons or organizations of interest, extracting confessions from political prisoners, recruiting additional agents, and/or carrying out paramilitary activities (Cohen, 2010: 153). Collaborators serve two supplemental roles that are more significant than their formal roles, functioning as a means of control to encourage "correct" conduct. The prospect that anybody could be a collaborator becomes an effective tool to encourage people to act and even think in ways that reduce organized dissent against the occupying powers. They also fragment Palestinian society, undercutting the confidence and trust required to build alliances and solidarity, and ignite political resistance (Gordon, 2008: 48). Our interviews and many informal conversations with residents in the West Line villages revealed that many locals believe that Palestinians granted permission to work in Israel, particularly *Alte Zachen* collectors (from the Yiddish phrase, which literally translates to "old stuff") and large scrapyard collectors (see Chapter 2), have been pressured by and made arrangements with the Israeli military to serve as an informant in exchange for permission to work in Israel.

1.1.2 Rise of an Informal Economy

The de-development of the Palestinian economy, the instability and high risk faced by formal enterprises, and endemic corruption have all created an economy in which much (and in some sectors, most) of the economy is informal. While the measurement of an informal sector is, almost by definition, difficult, it has been estimated that informal businesses in Palestine employ approximately 180,000 people, generate $300 million annually (Massar Associates, 2003), and contribute up to 50 percent of the national economy (Kawasmi & White, 2010). Informal businesses are often small, requiring little to no capital, provide moderate to low incomes and unstable employment, and frequently operate in unsafe working conditions. While the informal sector performs a valuable economic role in Palestine's economy, it has also obvious drawbacks, such as the vulnerabilities and lack of protection of workers and their families, and the undermining of tax collection and regulation (Hilal et al., 2008).

1.1.3 Spatial Fragmentation and Border Blurring

Israel and the West Bank share a unique border between the two territories, entailing a massive (but incomplete) separation wall, thousands of formal workers crossing the border daily (as well as considerable informal crossings), and a complex shifting geography of military occupation (see Figures 1.1 and 1.6). Under the current Oslo Agreement, stringent spatial controls have divided the West Bank into geopolitical Areas A, B, and C. Within this system, Area A is under full Palestinian civil and military control, Area B is under full Palestinian civil control and joint Israeli-Palestinian military control, and Area C is under full Israeli civil and military control. According to the Oslo Agreement, these demarcations define clear governance divisions; however, in practice, the borders between Areas A, B, and C are opaque at best. The border is selectively porous, and mobility differential. Thus, Israelis in settlements can move relatively easily through a system of bypass roads that emerged in the West Bank, and pass through borders in either direction with ease. At the same time, Palestinians face physical barricades and checkpoints as well as the "separation wall," and passage to Israel is highly regulated with severe penalties for

Figure 1.1 Photograph taken in the West Line village of Beit Awwa in 2016, showing the placement of concrete slabs to replace the chain-linked fence separation barrier. Before this, the concrete separation wall from the north of Beit Awwa was met by an easily penetrable chain-linked fence extending to the South.

infringement. The Palestinian Authority is not allowed to regulate incoming goods, while Israeli regulation in both directions is confounded by settlements in Palestine that are considered "Israeli" destinations so that e-waste loads might be coming from or to these. The selectivity of mobility and passage has encouraged corruption, smuggling, and an economy of checkpoints (Tawil-Souri, 2009).

1.2 The Rise of the West Line E-Waste Industry

Against this background, over a short period from around 2003 to 2006, the West Line villages evolved from hosting a scattering of e-waste scrapyards, as many Palestinian villages do, to becoming a centralized preeminent e-waste hub (see Figure 1.2). The hub's emergence can usefully be demarcated as 2004, coinciding with the heightened physical separation of the West Line villages from their

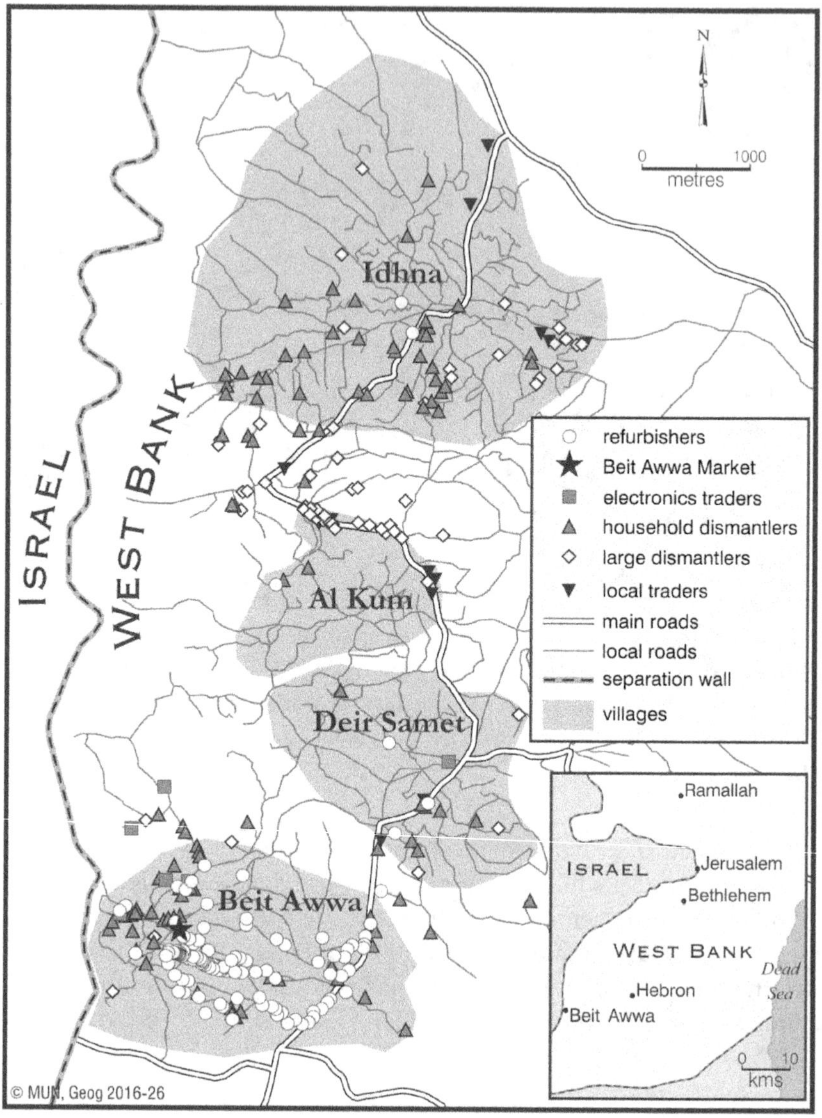

Figure 1.2 Map illustrating the clustered location of e-waste-dependent businesses in the West Line villages (Davis & Garb, 2018a).

neighbors and the skyrocketing metal prices of the commodity boom of the 2000s. This milestone is marked distinctly in the appearance of extensive e-waste burning – an indication of increased e-waste recycling activity – in our satellite imagery time-series records (see

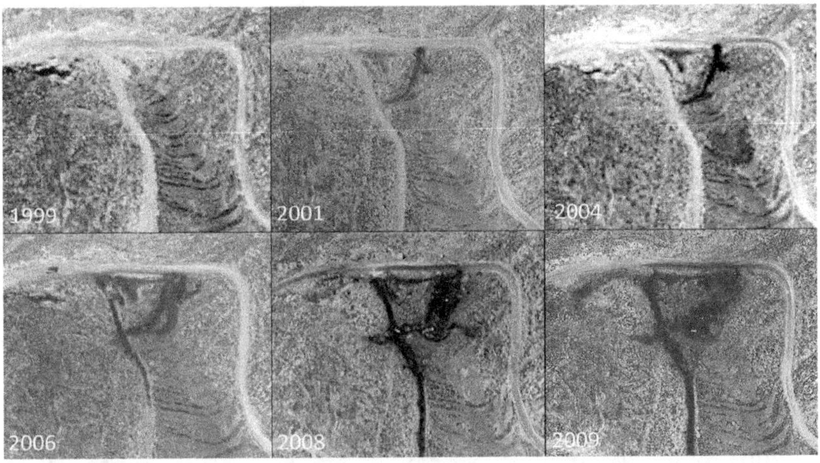

Figure 1.3 Time series of satellite images from 1999 to 2009 showing the progression of a primary site for e-waste burning on the outskirts of the West Line villages, which coincides with the progression of the West Line e-waste industry.

Figure 1.3).[5] This process, which was intended and propelled by those actively working in the industry, certainly was not state-planned; in fact, it remained seemingly invisible, garnering little attention from national Palestinian and Israeli governments for decades. How and why did this hub emerge, and why here and at this moment? Clearly, the broad-stroke parameters of income and regulatory differentials that prevailed throughout the West Bank for many decades are necessary but not sufficient explanatory factors. Here, we describe five additional attributes that primed the West Line villages to emerge as Palestine's main e-waste hub: (1) livelihood disruption; (2) proximity to Israel; (3) predisposing industry; (4) amenable social structure; and (5) weak environmental protection. These five attributes emerged as the most common responses during interviews with over twenty-five

[5] As part of a study to determine the prevalence and severity of burn sites in the Hebron Governorate, we systematically mapped the location and area of all e-waste burn sites from 1999 to 2009 using satellite imagery. The result of this study demonstrated a strong spatial association between a previously identified cluster of childhood lymphoma and a concentrated area of e-waste burn sites in the West Line villages (Davis & Garb, 2018b).

scrapyard owners and many more residents in the West Line villages, as we regularly posed the "why here?" question during our time in the villages. These attributes build on initial understandings of global and national e-waste flows that rely on the PHH and offer a more nuanced picture of the forces that shape cross-border e-waste flows and help recast e-waste destinations as not simply "digital dumps" but specialized industrial clusters.

1.2.1 Livelihood Disruption

E-waste recycling is a demanding and sometimes precarious form of livelihood, but a compelling income source where few other employment opportunities are available. When work in Israel was interrupted in 2001, many Palestinians faced desperate deprivation, but the West Line inhabitants were primed to seize upon a waste economy as an unlikely but lucrative alternative. The reduction in work permits following the Second Intifada had a strong negative impact on employment in the West Bank: Unemployment rates of 7.5 percent rose rapidly, reaching 28.2 percent in 2002 (Miaari et al., 2014). Increased unemployment mostly affected uneducated workers employed in Israel who received significantly higher wages than those available in the West Bank (Miaari & Saier, 2011). Because of the West Line's proximity and short commute to Israel, on the one hand, and distance from Palestinian urban centers on the other, they were particularly dependent on employment in Israel and had minimal investments and growth in local businesses.

This rise in unemployment acted as a catalyst for residents to search for any form of alternative employment, either long-term or temporary, which did not depend on permission to enter Israel. At the same time, international metal prices commonly found in e-waste rose to unprecedented levels from 2000–2008, spurred primarily by Chinese demand. For example, aluminum prices rose from \$1,400/tonne in 2002 to \$3,200/tonne in 2008, and copper spiked from \$1,800/tonne in 2000 to \$9,000/tonne in 2008. The existing relationships and historical roots in processing used furniture, coupled with an awareness of the growing value of raw metals that compose e-waste, fostered the surge of newly unemployed residents to shift from an economy dependent on Israeli employment to an economy dependent on Israeli e-waste.

Thus, a loss of usual livelihood source coincided with the emergence of a newly valuable resource, which could be easily and almost freely harvested. The low entry barriers to the waste economy make it especially appealing. And, as opposed to employment in Israel, which required all participants to enter Israel physically, here was an opportunity through which a few people crossing the border could leverage much wider circles of employment. Thus, a small number of Palestinians, who speak Hebrew and hold permits to enter Israel (and the financial capacity to hire an Arab-Israeli driver), can transport materials into the West Bank as a source of employment for thousands in the West Line who repair or dismantle these items or profit from the creation of supplemental businesses and the overall influx of wealth into the villages. The informal nature of the scrap industry, devoid of taxation and regulation, offers a considerable advantage relative to other forms of employment. Nor does the industry require advanced education, so it offers an easy transition for those who worked blue-collar jobs in Israel.

Thus, amid the general contraction of opportunities and incomes, the West Line found an inventive alternative. In fact, these new circumstances afforded some entrepreneurs a way to move from low-salaried and uncertain physical labor to relative prosperity. The combination of first-world revenues in Israel with third-world living costs in rural Palestine and the relatively low entry costs in the recycling industry enabled the more enterprising to accumulate sufficient capital to invest in their own business and incrementally increase the scale of operations. However, the majority, especially those who did not enjoy mobility across the border, worked in smaller dismantling and refurbishment businesses, or even household-level ones that can take place at home or on the street, with minimal overheads. Using basic hand tools (e.g., axe, hammer, screwdriver/power drill), an initial load of purchased e-waste can be dismantled and separated into metal components and sold to local traders. As the business builds up, profits can be reinvested to purchase higher quality and quantities of e-waste and more advanced dismantling machinery. Though work conditions are physically demanding and can hold many immediate and long-term health risks, employees of dismantling scrapyards can earn locally competitive salaries. Thus, thousands of uneducated Palestinians who lost their ability to work in Israel after the Second Intifada created a viable alternative in the context of deprivation.

1.2.2 Proximity to E-Waste Source

Global movements of e-waste typically require freight shipping over long distances to arrive at port cities in developing countries, which explains why most e-waste hubs that rely on international e-waste shipments, from both the Global North and Global South, are located near international ports (Greenpeace, 2009). However, the proximity and extensive road networks available to e-waste importers and exporters in the Israel–West Bank case make e-waste collection via trucks feasible, as most of the Israeli population lives within an hour drive from the West Bank border. The West Line villages are situated immediately adjacent to the Israeli border. Their location not only habituated an entire generation of local men to easy movement across the border over the decades of free flow, but still allows easy access for collection trucks to enter Israel in the morning, collect e-waste during the day, and return to the West Line villages in the evening. Within a one- to two-hour drive, collection drivers can readily service major Israeli cities – such as Tel Aviv, Ashdod, Jerusalem, or Beer Sheva – where dense populations hold higher quantities of e-waste. Though entry permits to Israel are typically given on a daily basis (disallowing overnight stays), collection drivers from the West Line can still make daily collection circuits and return to unload by evening.

Villages located even an hour's drive east or south of the West Line villages would unlikely be able to sustain this economy. The added cost of gas alone, which in the West Bank is among the most expensive in the world, would eat into the thin profit margins of e-waste collectors. Logistically, collection drivers from the West Line typically begin their day before six am and often do not return until seven pm. If collectors had to drive an extra hour to and from Israel every day, a challenging thirteen-hour workday would become an unbearable fifteen-hour slog. Thus, the range of locations that could host an e-waste hub in the West Bank is restricted to regions located within a short drive to the Israeli border crossing.

1.2.3 Predisposing Industry

E-waste hubs require a predisposing infrastructure of people, relationships, and knowledge that enables waste entrepreneurs to become aware of the value and profit from discarded materials. Before e-waste recycling dominated the West Line economy, these villages had a long history, dating back to the pre-state period, collecting and refurbishing used furniture and other household goods collected in Israel. This industry was supported by a profession of collection drivers that clustered in the West Line but were also prevalent throughout the West Bank and Israel, known as *Alte Zachen* (described in Chapter 2). Collected goods were transported to a vibrant auction in Beit Awwa, where refurbishers and families inspected newly collected goods and haggled over prices. Purchased goods were repaired and resold to Palestinians across the West Bank, putting Beit Awwa on the map as the place to buy inexpensive secondhand furniture and household items.

The Beit Awwa auction has served as a key node and the lifeblood of this industry, with sixty to seventy trucks arriving daily, carrying one to three tonnes of mixed loads of e-waste and used materials per truck (see Figure 1.4). This convenient centralized meeting place – the largest of its kind throughout the West Bank – has benefited both buyers and

Figure 1.4 An *Alte Zachen* truck pulling into the Beit Awwa Market. Already, prospective buyers are eagerly inspecting the quality of items to assess their value. The lively auction is just beginning to draw a crowd.

sellers. *Alte Zachen* collectors can negotiate high selling prices from a competitive auction, while e-waste buyers can find niche items (LCD TVs, printers, treadmills, washing machines, etc.) fostering specialization, which increases the potential value extracted from secondhand electronics. Were it not for this nightly auction, the dispersal of Israeli e-waste in the West Bank would be costlier and less efficient. Moreover, e-waste collection would be less profitable, driving collectors to another profession, reducing flows of Israeli e-waste.

Over time, electronic and electrical goods in all shapes and sizes became a greater, and more valuable, segment of the used products available to *Alte Zachen* drivers. Much of the increased availability of e-waste can be attributed to rapid advances in technology that have increased the presence of electrical and electronic goods in households, the rapidity of turnover, and the absence of any formal collection mechanisms in Israel, (E-waste was only belatedly recognized as an important policy issue around 2012.) As e-waste grew, much of the infrastructure and many of the business models were already in place for an easy transition from refurbishing furniture and household products to refurbishing and dismantling electronics.

Another, less traditional, form of predisposing infrastructure is apparent in the abundance of ad hoc landfills and disposal burn sites. The inexpensive disposal of non-valuable e-waste components is an economically advantageous externality to reduce e-waste processing costs. Because Israel sharply restricts Palestinian development in the area, rural land on the outskirts of these villages has remained largely vacant. Because the West Line villages are situated against the separation barrier to their west, the extensive Area C areas in the gap between them and the separation barrier have become even more of a liminal area that attracts dumping and burning. With minimal opportunities for Palestinians to reside on or productively monetize their land, many have used it to dispose of their e-waste remainders or profit from renting out their land as an informal landfill and/or burn site for other scrapyards. Land in Area C with absentee owners has become communal spaces for extraction and disposal burns and has even fostered collective maintenance norms (see Figure 1.5). For example, excessive burning in a burn site will eventually create a thick layer of loose toxic ash that blows in the wind and kicks up with foot traffic. When ash has piled too high, regular users of the burn site will pool finances to hire a backhoe or an excavator to scrap the top layer

Figure 1.5 This commonly used burn site located near a less fortified portion of the separation wall, comprised of a chain-linked fence, is regularly maintained by a collective of e-waste burners who push the loose ash accumulated from excessive burning to the side of the burn site, clearing the passage of trucks to continue burning e-waste materials here.

of ash to the edges of the burn site, maintaining a firm surface. Even the long iron rods used to spread burning cables to speed up the burning process are typically left at burn sites as a courtesy for the next burner.

1.2.4 Amenable Social Structure

A less tangible quality, yet nonetheless influential, is a social structure with great cohesion. In the case of the West Line, family-based clans offer a high trust setting needed to operate an informal business and semilegal economy in which the protection of industry information, such as e-waste buying and selling sources, is imperative. Simultaneously, the clan-based occupational choice to work in e-waste helps contain and neutralize the stigma of working in a "rubbish" economy and neutralizes opposition and protest against its harms.

While financial investments present a low entry barrier into the e-waste industry, there are high barriers of contacts, skills, and tacit knowledge, which are gradually acquired through industry immersion, and often closely guarded. Only through experience can one learn the intricacies of the industry: where to find the most valuable e-waste for the lowest price, who to sell various dismantled components to, how to estimate the value of mixed scrap piles, how to negotiate the changing value of e-waste items, or how to economically dismantle various e-waste goods. One cannot get a degree in scrap recycling but must learn it directly, which explains what some villagers have described as the "contagious" nature of the e-waste profession. The opportunities of the e-waste trade can only be experienced through personal connections.

These dynamics are exemplified in the rise of the circuit board subeconomy in the West Line. Knowledge of the value and markets of circuit boards came initially from two nearby Israeli settlements, Adora and Telem. Before the emergence of the West Line e-waste economy, there was little perceived value in circuit boards in the few scrapyards that dismantled e-waste in the 1990s and early 2000s. For the most part, the focus was on extracting copper, aluminum, steel, and iron, and circuit boards were typically stripped of these metals and then discarded. In 2004, an Israeli settler shared with a scrap dealer from Beit Awwa the true value of circuit boards, which contain gold, silver, palladium, and other precious metals. Because this knowledge was well guarded, this scrap dealer benefited from high profit margins between the high-selling-price circuit boards he received from the Israeli buyer and the low price at which he purchased from dismantlers, whom, until this point, had been discarding these and welcomed this extra revenue stream. Collected circuit boards were brought back to a warehouse and organized based on specifications required by foreign refineries. The categorized circuit boards would then be sold to one of two Israeli settlers who functioned as middlemen by facilitating their transportation from the West Bank into Israel to avoid taxes and difficulties with customs. The circuit boards would then be sold to European refineries for metal extraction. After maintaining a monopoly in this new subeconomy for over three years, two additional relatives of this Palestinian pioneer became familiar enough with the purchase and selling prices and made independent connections with the Israeli settlers to open their own competitive operations. This new competition more than doubled the purchase price of circuit boards, to the benefit of dismantling facilities and the consternation of the original circuit board dealer.

The tendency of e-waste to be a family business also allows the profession to become established despite prevailing stigmas against this unglamorous form of livelihood, which is usually inherently dirty and physical, leaving its practitioners with perpetually stained black hands from handling scrap metal – a telltale sign that is not easy to hide. Mothers will steer their daughters away from marrying into families with reputations for working in the scrap industry. But since marriage is also very much along family lines, once a family has entered this trade, the costs of this stigma are neutralized. We can also observe how participation in the scrap sector, and even particular specializations within this, is dominated by specific families. About half of the dozen prominent extended families in the West Line villages work in the sector. Beit Awwa's population of approximately 8,000 residents, for example, has consisted of two extended families for over 100 years: the Masalma family living in the east and the Sweity family in the west. The Masalma family dominates the refurbishment sector, while the Sweity family makes up most of the *Alte Zachen* profession and hosts many small dismantling facilities. In the neighboring village of Idhna, the Batran, Abu-Zalata, and particularly the Abu-Jheshi families are well-known for operating the larger dismantling scrap-yards and local metal traders (see Figure 1.2).

Alternatively, certain families within the West Line villages abstain from the e-waste industry altogether. The Al'Joub family, numbering over a thousand, live in the heart of the West Line villages in Al Kum adjacent to the main strip of scrap yards. They are adamantly against this industry's dirty nature and the pollution it causes, preferring to work in construction and agricultural professions, even though they earn a more modest income.

This family-based social structure that modulates participation in the industry between and within family groups has also defused opposition to it for decades, even as its harms became more salient. While confrontations still erupt over air pollution from e-waste burning, particularly when burning occurs close to residential areas and poisons infants and small children, the village's overall economic dependence on e-waste has constrained community pushback. The balance between harm and livelihood is unevenly felt in different clans, with some predominantly experiencing the harm side of the equation, and others more ambivalent, since, even if the immediate family does not work in the trade, many in their extended family do. These dynamics surface in the rare and ineffective attempts to punish e-waste burners. Typically, when e-waste burners are arrested by either the local municipality, police, or Palestinian Environmental Quality Authority (EQA), the burner receives a minor fine, sometimes as low as ₪200 ($58), and signs a pledge promising they will not burn again. Traditional tribal hierarchies and modes of conflict resolution are very much alive, so that respected heads of the families filing complaints and those creating the problems will meet. The latter will make a plea for the economic hardships of the e-waste burner, their impoverished living conditions, and the absence of viable alternative livelihoods – not to mention the internal family hostility that would ensue for years were severe penalties applied. These discussions typically result in the complaint being withdrawn and the charges dropped. These kinds of conversations extend to the municipalities, and far beyond, to the police, and the regional EQA officials. Thus, family cohesion has both facilitated entry and blunted opposition to the e-waste industry.

1.2.5 Weak Environmental Protection

Beyond the general differential between environmental enforcement and regulation in Israel and the West Bank, the West Line villages "enjoyed" an additional and more local governance vacuum. Under the Oslo Accords, the residential areas of the West Line villages were designated as Area B, permitting Palestinian civil control, whereas land surrounding residential areas was designated as Area C, where the Israeli military hold jurisdiction and place restrictions on Palestinian policing (see Figure 1.6). These conditions have left environmental protection in disarray due to this responsibility being ineffectively shared between understaffed and jurisdictionally constrained Palestinian police and Israeli military that place a low priority on environmental protection.

The interplay of the e-waste industry and the West Line villages' geopolitical divisions surfaces in the strategic location of scrapyards and burn sites. Most scrapyards are in Area B since scrapyards built in the development-restricted Area C are typically demolished by the Israeli military. At the same time, most open burning historically occurred in Area C, which is a governance vacuum as Palestinian police cannot access this geopolitical region without advance Israeli permission (which can take up to forty-eight hours to obtain) while

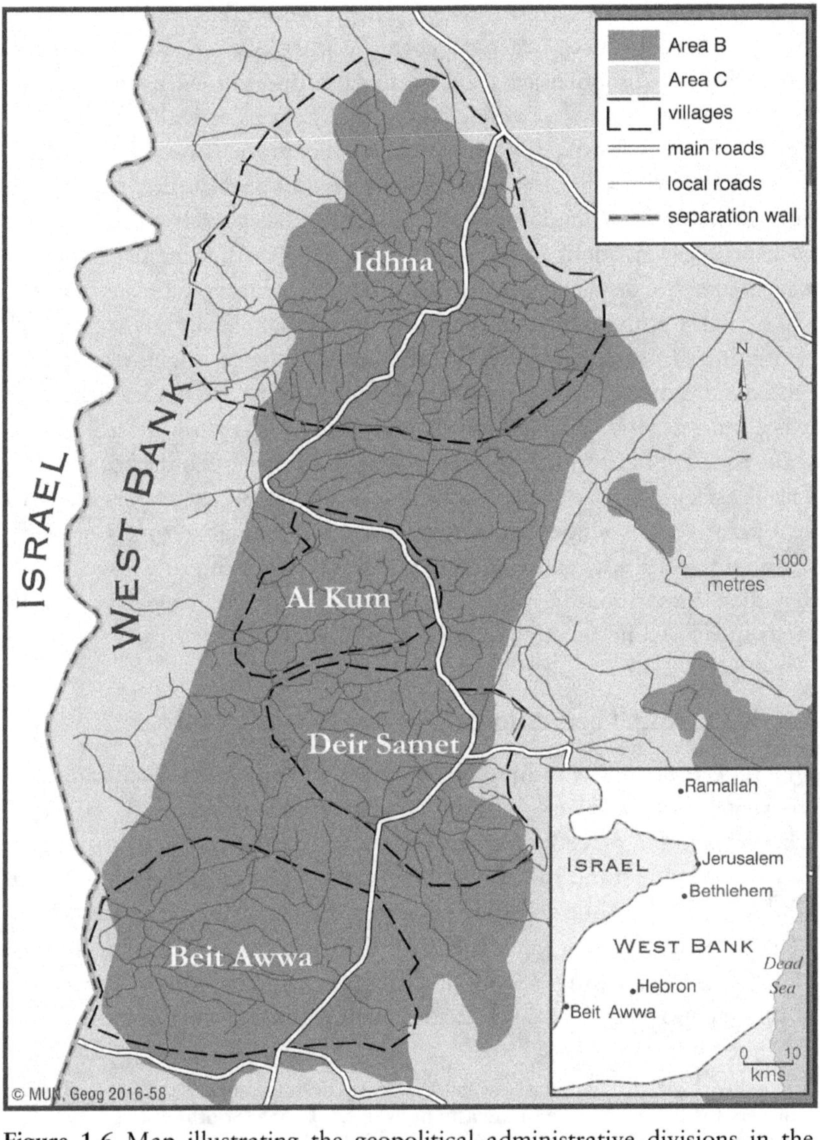

Figure 1.6 Map illustrating the geopolitical administrative divisions in the West Line villages. The West Bank is divided into three geopolitical regions: Area A (not shown on the map) designates land under full Palestinian civil and security control, Area B designates land under full Palestinian civil control and joint Israeli-Palestinian security control, and Area C designates land under full Israeli control for security, planning, and construction.

Israeli military forces are rarely present in these areas, and when they are, they tend to have other concerns. Within Area C, the larger and more continuous burn sites gravitate to properties with absentee or municipal ownership, or very close to the separation wall where interference from Palestinian authorities is even more constrained. Thus, the Oslo divisions of space were conducive to the West Line e-waste hub: availability of land in Area C that could be severely contaminated with little government interference immediately adjacent to Area B in which scrapyards could operate with little regulatory interference.

The PHH emphasizes national-scale discrepancies in environmental regulation and enforcement as the driving force for international e-waste practices. Yet, while regulation is laxer in places where e-waste hubs are found, this is hardly an explanatory factor at the national scale, as it applies to many places in which e-waste hubs do not emerge. The West Line e-waste hub shows that a regulatory vacuum is likely to be operative not simply at a national scale, but at a much smaller, regional scale. Reduced ownership and governance offering a weak link in which contaminating processes can operate unchallenged prevail in areas measured in hectares or square kilometers.

1.3 Recurring Themes in Analogous E-Waste Hubs

Our reflection on the unique place-based attributes that facilitated the emergence of the West Line e-waste hub prompted us to consider whether similar place-based attributes (and possibly others) might be helpful to understand the patchy emergence of e-waste hubs across the Global South.[6] While the particulars of the emerging situation of the West Line are unique, key aspects of the motivations for and absence of barriers to harmful recycling practices are common to the transboundary movement of e-waste to hubs throughout the Global South.[7]

[6] The theoretical discussion of e-waste flows and destinations, and the contributing drivers for the emergence of e-waste hubs in the Global South, draws heavily from our previous publication in *Geoforum* (Davis et al., 2019).

[7] There is much debate on the destination of international e-waste flows, and, particularly, about what percentage of e-waste from the Global North arrives in e-waste hubs in the Global South and how much e-waste that is processed in these hubs are from domestic sources. These debates stem from challenges quantifying e-waste flows due to a lack of consistent definitions for waste materials and no real trade data that captures formal e-waste flows, and especially not informal flows. To date, methods to track e-waste have relied on small-sample Global

In the Israel–Palestine case, the discrepancies in labor costs, environmental regulations, and enforcement in peripheral areas are intensified by their geographic proximity and the porousness of the border, and it is also one of the few, perhaps only, places where there is almost exclusive one-way transfer between e-waste importers and exporters (i.e., almost all Palestinian inputs come from Israel, and most Israeli outputs flow to the Palestinian areas). The PHH offers a macro-level theoretical skeleton to understand e-waste flows, but alone is an insufficient predictor. We suspected that the "why here?" question could show the dynamics beneath the thin accounts provided by NGO reports and media exposés, which focused on the poor working conditions and excessive pollution on the receiving end, and overconsumption and irresponsible e-waste management in the Global North on the other (*60 Minutes*, 2008; *Frontline*, 2009; Puckett et al., 2002; *USNEWS*, 2014). Indeed, beyond the sweeping explanations of these places as victimized global "digital dumpsites" for the refuse of rich countries, a more complex story is apparent, with striking similarities to the West Line case. To date, there has been limited in-depth research exploring the origins and emergence of e-waste hubs in other sites, and we suggest future research addressing this question might reveal similar patterns as we found in the West Line villages. However, there are hints in the literature of common origin stories relating to some of the five factors we identified. Our review (Davis et al., 2019) of two well-known sites, the notorious Guiyu hub in China and the Agbogbloshie hub in Ghana, shows several compelling commonalities that have been discussed by other researchers in the economic, governance, and social conditions in which these e-waste hubs emerged.

Guiyu, which was the focus of the early and widely publicized exposé *Exporting Harm* by the Basel Action Network (BAN) (Puckett et al., 2002), has become the poster child for discourses of irresponsible e-waste exportation from the Global North. Many accounts point to discrepancies in environmental protection and

Positioning System (GPS) tracking of e-waste items (Lee et al., 2018), anecdotal reports (Breivik et al., 2014; Efthymiou et al., 2016; Puckett et al., 2002), or proxies for e-waste (Lepawsky & McNabb, 2010).

worker salaries as the reason e-waste is directed here, but a closer look at the history of the regional economy shows that, as with the West Line villages, Guiyu had processed discarded materials since the 1950s, which became an entrenched and primary sector centered on e-waste after a disruption of primary livelihoods, in this case agriculture (Li et al., 2011). Due to Guiyu's location in a low-lying flood-prone area, poor agricultural production displaced farmers, especially in the northern provinces of Hunan and Anhui, who migrated to Guiyu (Puckett et al., 2002). Destitute and socially marginalized, they overcame reticence to work in the stigmatized scrap industry, where they provided the physical labor. As in the West Line, the availability of liminal unregulated areas (in this case the disarray of land administration in areas falling between the jurisdiction of two neighboring cities, Puning and Jieyang) offered a location where these contaminating activities could be established with less resistance (Li et al., 2011).

Agbogbloshie is another widely known e-waste hub. Here, too, ambiguous land ownership has played a critical role in facilitating the emergence of an e-waste hub, which emerged in the environmental governance vacuum of an area on which migrant settlers had squatted for decades, creating ownership ambiguity and impeding proposed development plans of the government of Ghana (Afenah, 2012; Grant, 2006). Here, the livelihood loss that served as a trigger for the emergence of an e-waste sector was more remote: the loss of income from rural agriculture in northern Ghana following declines in agricultural productivity and labor in northern Ghana due to 1980s Structural Adjustment Programs (SAPs) that reduced agricultural subsidies and government support for health, water, and education for many rural populations. While the SAPs exacerbated poverty in rural Ghana in general, the northern regions bore the brunt as agriculture shifted from food crops to export crops like cocoa and timber that were mostly grown in the south (Anyinam, 1994; Konadu Agyemang, 2000; Yeboah, 2000). With declining agricultural productivity, many rural dwellers from the north migrated to urban areas. Agbogbloshie offered, and continues to offer, cheap rent and easy entry into scrapyard work for many migrants seeking better opportunities through informal waged labor, especially from the northern part of the country, which accounts for over 70 percent of these workers (Amankwaah, 2013; Prakash, 2010). Here, too, migrant workers and marginalized populations were less deterred by the stigmatization of working with rubbish.

1.4 The Book

This chapter has tracked the emergence of e-waste hubs historically, as places taking shape within and because of specific local circumstances. The upshot of the kind of more detailed accounts we have offered here is not only a more nuanced political geography of e-waste hubs, but a reframing of their agency and functioning and, therefore, of the necessary policy stances and measures for reform. Hubs cannot be seen simply as victimized and passive receptors for the dumping of wastes mobilized by national-scale inequalities, as conceived by the PHH. They appear as the result of active entrepreneurs who have labored to forge and stabilize systems of collection, transport, processing, and profit extraction that transform waste into a rare source of income in a harsh economic landscape.

The remainder of this book is divided into two sections: the first explores the functioning, portrayals, and futures of e-waste hubs and the second a potential pathway for their reform, and their role in reforming e-waste policies more broadly. The remaining chapters in the first section (Chapters 2–5) describe the economic and environmental consequences of the informal e-waste economy in which e-waste hubs such as the West Line are central. Yet the portrayals of these sites that have dominated perceptions and policies regarding e-waste treatment have been Global North–centric. That is, they imagine a landscape without a robust and longstanding informal sector, neglect the importance of livelihoods, and aspire for e-waste management that is centralized and generates "new" green economies with state-of-the-art recycling technology. The result is an exclusive focus on remedying future environmental harm with profitable waste management business models while eclipsing the environmental harm of past e-waste along with the vitality and active economic roles of informal recyclers. We describe our community engagement method in the West Line villages, in which a hub-centric stance helped elicit and elaborate a different set of proposed responses to the situation, reflecting the balancing of hazards and opportunities of e-waste hubs from those most affected – the residents of the sites in which most of the world's e-waste is processed.

While Chapter 1 describes the emergence of the West Line e-waste hub, Chapter 2 further sets the scene of this book through an in-depth description of how it functions and the economic and health impacts of the scrap sector. Our description of the e-waste economy proceeds sequentially from collection in Israel, the smuggling of scrap across the border to the West Bank, the recycling process and markets in the West Line villages, and the resale of extracted materials back to Israel. Because the scope and seriousness of this cross-border economy remained fairly invisible to NGOs, media outlets, and Israeli and Palestinian national government authorities for well over a decade, and since there is a lack of detailed descriptions of the entire economic landscape of e-waste hubs more generally, we describe this context in some detail, contextualizing it within the uneven power dynamics of the West Bank.

Chapter 3 elaborates on the context and consequences of the dominant portrayals of e-waste hubs as "digital dumpsites" for the Global North. We argue that this framing has led to narrow portrayals and crude policies that attempt to remove e-waste from these sites through a combination of e-waste trade bans and EPR-based e-waste management regulation that establish new formal e-waste recycling systems. In the Israel–Palestine context, the West Line e-waste hub became visible to environmental NGOs through selective engagement with a minority of West Line residents who are hostile to the e-waste economy and able to reach out to influential audiences. These environmental NGOs situated this case of e-waste contamination within a broader "digital dumping" narrative, which was then publicized and shared with Israeli and Palestinian government authorities. While Israeli and Palestinian authorities held diametrically opposing views of where the blame lay, they initially, somewhat ironically, urged quite similar policy responses to cut off all e-waste flows from Israel to this e-waste hub – a blunt approach that, if effective, would displace livelihoods without remedying the enduring toxic legacy in the West Line villages.

Chapter 4 wrestles with the complexity and challenges of accessing and engaging a more representative range of actors who might forge a consensual trajectory to reform harmful practices and retain livelihoods. We explore the diversity of stakeholders and perspectives within e-waste hubs, as well as the ironies of our own role as outsiders working to allow such community visions to emerge through an unconventional community engagement method that ensures consideration of this diversity. We unpack our considerations and the limitations of this community engagement approach, which entailed an initial phase of mapping the relevant stakeholders and community cleavages, and then deploying a novel Delphi-like method with eight "community representatives" through a spoke and hub form of communication between them and the facilitator that enabled us to forward a broadly endorsed development trajectory within a heterogeneous and conflicted community.

Chapter 5 broadens our process within the West Line villages to point to a hub-centered approach to the global e-waste problematic, that is, one in which the agency and interests of the places where e-waste is primarily processed figure large. We describe why a hub-centered approach offers a strategically, politically, and ethically advantageous leverage point for the desired changes. Drawing on the West Line villages' desired development trajectory outlined in the previous chapter, we argue for the necessity of a policy package to reform e-waste hubs that consist of three pillars: eliminating environmentally harmful recycling practices, remediating past contamination, and opening pathways to reform and regulate existing e-waste hubs. These pillars led into the second section of the book and structure the following three chapters.

In the second section (Chapters 6–9), *Pathways and Predicaments*, our attention shifts to our activities in advancing the hub-centered proposals for reform that emerged from our work in the West Line villages, and what we learn from them about e-waste sector reform more generally along with the Israeli-Palestinian geopolitical setting. The hardships and friction we encountered in pursuing locally defined solutions uncover the paralyzing political dynamics *between* Israel and Palestine and *within* the Palestinian Authority to remedy cross-border environmental issues and offer a direction and process to reform e-waste hubs throughout the Global South.

Chapter 6 unravels the complex dynamics of e-waste burning, as a method both to extract metals wrapped in plastic and to dispose of non-valuable discards. This burning has become the visible, indeed, iconic public face of e-waste hubs across the Global South, and we describe the grave human–environment consequences of this practice in other e-waste hubs, along with our own findings in the West Line villages. We then explore the conditions that enable e-waste burning to persist undeterred, which, in turn, informed our strategy to eradicate it. During a three-month pilot project, we combined the carrot of subsidized cable grinding to offer an alternative to burning with the stick of enhanced enforcement through a grassroots community environmental policing system to overcome previous policing barriers. While the outcomes of our pilot project demonstrated a feasible and financially sustainable strategy to significantly reduce e-waste burning, the successful uptake in the Palestinian context has been constrained by unresolved national disputes over statehood and between central and local authorities within the Palestinian Authority.

In Chapter 7, we turn our attention to the neglected toxic legacies of e-waste hubs and bring them into conversation with prevalent e-waste management policies. As a point of departure, we review the human–environment burden of e-waste burn sites, highlighting the risks posed by the toxics generated at and dispersed from these sites. We argue that remediation must play a central role in a hub-based approach to e-waste management and discuss the donor-funded remediation plan we developed in the West Line villages. This chapter concludes by dissecting the political challenges and divergent agendas we encountered with Israeli and Palestinian authorities that derailed a carefully crafted plan to remediate e-waste burn sites in the West Line villages.

Chapter 8 details the strategies and advocacy efforts we forwarded to integrate the West Line e-waste hub with the recently formalized Israeli e-waste sector in a way that would allow both to benefit from a trajectory toward sustainability and upgrading of the West Line hub. We describe our engagement with key decision-makers on both the Israeli and Palestinian sides, the unexpected allies and adversaries of these efforts and their stances, and the significant legal, political, and logistic barriers that inhibit an equitable integration of these two systems. These tensions and dynamics in the Israel–Palestine case echo those that occur throughout debates surrounding management approaches to e-waste hubs in the Global South and Global North–Global South e-waste trade broadly, in places where colonial relations are less visible and immediate than they are in the Israeli-Palestinian context.

In the final chapter, we take stock of the current state of play and outlook of the West Line villages. We use this as a platform from which to consider the broader forces that inhibit the prospects of upgrading e-waste hubs in the Israel–Palestine context and beyond, and as a call to action for relevant stakeholders to support and facilitate hub-based reforms.

2 *The West Line E-Waste Economy*

A perfect storm of circumstances facilitated the emergence of certain regions in the Global South as e-waste hubs. For the West Line villages, the rapid growth in Israeli e-waste production far outstripped the recognition of this phenomenon as a problem or policies for managing it. For years, Palestinians from the West Line villages lacking local employment and income opportunities had been commuting across the surprisingly porous border between the West Bank and Israel, often bringing back discarded yet valuable items. While this 708 kilometer long border is widely perceived to present a highly regulated and guarded security barrier, it is still prone to significant illegal crossings (UN OCHA, 2011). As unemployed Palestinians discovered e-waste, collecting and trading these items became an attractive, novel source of income. At the same time, there was little to restrain the emergence of a recycling industry based on deleterious dismantling practices, especially in rural villages in the geopolitical Areas B and C. The decade-long stasis in the process aimed at achieving a peace treaty envisaged under the 1993/1995 Oslo Accords left the West Bank politically and geographically fragmented, with the areas surrounding the West Line villages in a governance vacuum with little restraint on practices such as waste disposal and burning.[1] In this way, the West Line hub emerged in parallel with other e-waste hubs throughout the Global South and, for nearly two decades, flourished with minimal resistance from Israeli and Palestinian national authorities.

Thus far, we have explored various triggers that drove certain regions to emerge as e-waste hubs. Understanding the place-based histories and socioeconomic, geographic, and political circumstances that attract and facilitate the emergence and agglomeration of e-waste

[1] This section builds on a briefer description of the West Line e-waste economy in an article "Extended responsibility or continued dis/articulation? Critical perspectives on electronic waste policies from the Israeli-Palestinian case" that was published in *Environment and Planning E: Nature and Space* (Davis & Garb, 2019).

hubs is critical to challenging and improving e-waste trade policies grounded in the Basel Convention that imagines e-waste from the Global North as being dumped in the Global South. Yet, the structural factors at the macro level, suggested by the pollution haven hypothesis, coupled with the meso- and micro-level factors and initial kick-off triggers suggested in the previous chapter, provide a static picture of hubs. The industrial clusters literature tells us what happens next as preexisting traits as well as serendipitous minor events enable certain regions to pull ahead in a path-dependent trajectory through which increasing returns and competitive advantage allow one area to lock in its status and stability, even as other potential areas fall away (Braunerhjelm & Feldman, 2008). E-waste hubs emerge and flourish after years of this accretionary virtuous growth cycle. Over time, the gradual accumulation of expertise and tacit knowledge, networks, complex supply chains, specialization, niche diversification and synergies, and auxiliary support systems (which, conversely, represent sunk costs and spatial lock-in) are all fed by increasing economies of scale, so that a hub becomes an actor and magnet for e-waste flows. This emergent and even entrepreneurial dynamic is quite different from the dumping ground/pollution haven imagery that guided early treatment of such hubs.

This chapter builds on our discussion of e-waste hub emergence to offer a thick description of the West Line e-waste industry as an emblematic hub with complex economic and environmental consequences. Scholars have begun to trace global destruction networks (GDNs) and labor as an intrinsic counterpart to global production networks (GPNs) (Herod et al., 2014; McGrath-Champ et al., 2015; Millington & Lawhon, 2019), challenging the traditional emphasis on production and consumption in global commodity chain/value chain and production network analyses, which largely ignores the "rubbish" that is the "shadow" side of production. In fact, as e-waste moves through spatial and social locations, it blurs the categories of product and waste, "commodity" and "non-commodity," and value, non-value, and even negative value. Global destruction networks begin when commodities reach their end-of-life in GPNs and serve as reverse supply chains by injecting materials back into GPNs when recycled materials are introduced to new manufacturing activities. By following post-consumption economic practices and fixating on the afterlives of commodities, waste scholars have begun to unravel and debate the "ongoingness" of economic activities related to the recycling and refurbishment networks of discarded materials (Brooks, 2013; Herod et al., 2014). Several geographers of waste have shown how discarded materials are remade into new products and/or stripped down to their raw materials to feed into new rounds of production, exchange, and consumption to create newly formed GPNs by recirculating value (Crang et al., 2013; Lepawsky & Mather, 2011). Others (Herod et al., 2014) sympathize with the ongoingness of waste, yet argue there is a limit to the reuse of commodities within geographically constrained spaces. The GDN analytical framework applied to e-waste economies has revealed how informal and small to medium-sized businesses, and their embedded sociocultural fabrics, tend to produce polycentric networks (McGrath-Champ et al., 2015; Wang et al., 2021), which differs from GPN frameworks that foreground the role of transnational corporations as a driving force behind the strategic coupling of networks and regional development (Coe & Yeung, 2019; Yeung, 2016). Thus, theorizing GDNs demands a grounded and bottom-up analysis of local communities, places, and actors and close attention to the intersecting spatial, sociocultural, and micro–political dynamics that shape actors' experiences and linkages within the system, which is decidedly different from the top-down regulatory and macro-regional economic planning focus of GPN scholars (Hudson, 2004; Inverardi-Ferri, 2021; Wang et al., 2021).

Our understanding and description of the West Line e-waste destruction network emerged from four years of deep immersion in this region, which comprised over 300 qualitative interviews with stakeholders working within the e-waste economy as well as Israeli and Palestinian government agencies and NGOs that support or influence the industry. This sustained and immersive research served as a valuable precursor to determine the geographic and institutional contours of the industry and the characteristics of e-waste businesses. Our reconnaissance period informed a structured business analysis survey conducted in 2015 where we identified 381 e-waste businesses segmented into six business types (electronic traders, local traders, large dismantlers, household dismantlers, *Alte Zachen* collectors, and refurbishers), and conducted seventy surveys with a representative sample of each business type to quantify the sources, volumes, and values of e-waste materials entering the hub, and the sector's overall contribution to the Palestinian economy. The survey was supplemented with interviews and observations to triangulate aggregated quantities and flows of e-waste reported by businesses. These methods included direct observation of truck passages and contents to provide data on the kinds and prices of items as well as the number of trucks entering the system. We also conducted semi-structured interviews with sellers and buyers of all metals collected from the West Line businesses to further aggregate sales information from the survey. (For a detailed overview of our methods, see Davis & Garb, 2018a.) Our description of the West Line e-waste industry is inspired by Cook's (2004) "follow the thing" approach and proceeds sequentially along the e-waste flow: beginning with collection in Israel, passage to the West Bank, processing and extraction of metals, and transport/sale of these back to Israel – and international destinations. This account highlights certain themes. As we follow the e-waste flow, we show how state military interventions driven by the occupation have shaped the structure of the e-waste economy and created power imbalances between Israeli and Palestinian actors within the system. We also describe the impressive expertise and agglomeration efficiencies of this industry, on the one hand, and the environmental and public health tragedies it has led to, on the other, and how these remained without substantive intervention from Israeli and Palestinian national authorities for so long.

Our description of the West Line e-waste economy seems to be the first that offers this kind of detailed portrayal of the contours and flows of one instance of e-waste trade and of the complex industrial ecosystem of the e-waste hub on the receiving end. Descriptions of Global North–Global South e-waste trade tend to trace general trends between countries (Breivik et al., 2014; Efthymiou et al., 2016; Lee et al., 2018; Lepawsky & McNabb, 2010; Puckett et al., 2002), with little consideration of the means through which recyclers in the Global South source e-waste in the Global North or the connections and logistics that shape how e-waste is imported and exported within an informal system.[2] The collection and recycling of domestic e-waste by the informal sector have received more attention (Grant & Oteng-Ababio, 2012; Rochman et al., 2017), but the focus is typically on hazardous dismantling processes or how informal recyclers remove e-waste from

[2] Wang et al.'s (2021) analysis of Guiyu offers a recent exception that unpacks the evolution and local dynamics of this e-waste hub.

formal systems. Our in-depth study of one regional system, in which there is a singular mapping of source onto destination, allows us a deeper and more holistic understanding of the multiplicity of actors involved in the e-waste destruction network. As we show in later chapters, this kind of textured understanding of the processes and actors also offers a valuable basis to consider e-waste management policy responses that center interventions and reform efforts within longstanding e-waste hubs.

2.1 Collection Pathways

In contrast to e-waste hubs in Asia and Africa, which process domestic e-waste along with foreign e-waste that arrives in shipping containers from distant countries, the West Line e-waste hub relies on an informal fleet of trucks to directly collect e-waste and used materials from all areas of Israel and the West Bank. Since the early 2000s, nearly 100 Palestinian-owned trucks collect daily 250–350 tonnes of e-waste and mixed scrap materials (including items for refurbishment and use), transporting them back to the West Line villages for processing (Davis & Garb, 2018). E-waste is an important component in these imports, and roughly 90 percent of the e-waste that arrives in these villages comes from Israel. Every morning, collection trucks fan out across the country, scouring households and organizations for e-waste or scrap metal, which is more plentiful and of higher value than Palestinian sources. Over the last decade, the mobility restrictions increasingly placed on the Palestinian population under the Israeli occupation have necessitated the facilitation of Arab residents of Israel to collect e-waste in Israel, as they can own and drive vehicles within Israel, and cross borders with greater freedom.

The arrangement works as follows. A minority of Palestinians can legally enter Israel on a semi-regular basis by virtue of obtaining a short-term entry permit, typically for the workday. They are prohibited from driving in Israel – a measure initiated after a spate of Palestinian terror attacks involving vehicles, which became a permanent restriction. Because Palestinian scrap dealers prefer to circulate and negotiate directly in Israel, this permanent driving ban created a vital niche for Arab drivers with Israeli residency (overwhelmingly from East Jerusalem) to help Palestinians on their collection rounds. These drivers commute about forty-five minutes each morning from Jerusalem to the

Idhna–Tarqumiya junction, located at the northern end of the West Line villages and about 500 meters east of the Israel–West Bank border crossing. Here, they meet their Palestinian partner driving a collection truck with Israeli license plates, which the partner with Israeli residence will then drive through into Israel after letting the Palestinian off just before the checkpoint, as they are required to enter Israel only through the pedestrian lane after presentation of their permit and security check. Reunited on the Israeli side, both continue to their day of scrap collection within Israel. At the day's end, the loaded collection trucks arrive back at the Idhna–Tarqumiya junction, where the Arab-Israeli driver is dropped off at his car to commute back to Jerusalem, earning around ₪400 ($115) for the day's work. The driver's role is primarily driving unless the Palestinian party loses their permission to enter Israel, in which case they can, for a short while, act as a remote proxy on behalf of the Palestinian partner in these transactions. A scrap collector or trader that loses their entry permit for any sustained period will rapidly need to move "downward" to less lucrative levels of the e-waste system (such as domestic collection, small-scale dismantling/refurbishing).

The network and mechanisms for collection in Israel have been carefully established over many years. While these may seem chaotic to an outsider or Israeli authorities, there are four broad strategic collection sources, each attracting specific types of collectors with different business models. The most visible are collection trucks that search residential streets, publicly advertising over a loudspeaker that they are collecting and paying for all types of scrap, including e-waste, used furniture, auto components, or any material that can be resold or dismantled for a profit. Known in Israel as *Alte Zachen*, these collectors tend to have a uniform, semi-open three-tonne truck – small enough to maneuver on narrow residential roads but large enough to haul one to three tonnes of scrap. This kind of service has operated continuously in evolving formats for almost a century, dating back to the pre-state period. The *Alte Zachen* collectors from the West Line villages join similar trucks originating from other regions in the West Bank as well as Arab-Israeli collectors who sell their treasures to scrapyards or at second-hand markets in Israel.

Alte Zachen collectors have developed an impressive collection network extending to all major Israeli cities. Indeed, they have long achieved the kind of comprehensive household collection capacities that Israel's e-waste Extended Producer Responsibility (EPR) policy now aspires to achieve (discussed further in Chapter 4), efficiently giving a second life to e-waste that would otherwise most likely end up in landfill. These collectors have built up recognized collection territories, whose infringement can be met with hostile confrontations. E-waste and scrap metal from residential sources mostly consist of items already on the pavement, supplemented by larger items of residents responding to their calls. Israeli residents will typically flag down the driver and pass on these items for a small payment or, most often, for free in exchange for the removal service. This simple and flexible service makes the *Alte Zachen* a familiar feature of Israeli towns, especially those within easier reach of the West Line villages. In Beer Sheva, for example, an old TV or refrigerator set out in the morning will seldom remain in place for more than a few hours.

This direct collection of occasional items from households with limited knowledge of scrap values offers one of the higher profit margins in the e-waste economy. An air conditioner, for example, is one of the more common and lucrative items, and an *Alte Zachen* collector who offers to haul one away for free, or even pays ₪100 ($29) for it, can sell it in the West Line villages for the parts and materials alone for ₪400–500 ($115–144); if the machine can be refurbished, it can sell for ₪1,000–2,000 ($287–574). *Alte Zachen* collectors are also well positioned to augment this steady collection of household items with ad hoc opportunities, such as scrap metal found along the side of the road, or special collector's items, such as vintage radios or typewriters.[3] While *Alte Zachen* are the most visible and recognizable collectors throughout Israel, they convey a variety of used materials (e.g., furniture, scrap metal, and household items) for resale. Within

[3] As we discovered the value of discarded electronics and used machinery, we were inspired to partake in the *Alte Zachen* economy and, in the process, witnessed their impressive ability to take advantage of impromptu opportunities. For example, researchers in our academic department were discarding a large experimental brick-making machine used for architectural research, so we offered to find someone who could remove it from the university free of charge. We arranged for collection by an *Alte Zachen* friend from the West Line villages. However, it turned out that this was not his most profitable transaction while on campus. He disappeared for about thirty minutes while the brick-making machine was being loaded, returning with the facility manager in a golf cart stacked with four discarded air conditioners. These would earn him more than twice the bulky brick-making machine that originally brought him to the campus.

the West Line villages, e-waste comprises only 5,700 tonnes per year out of over 37,000 tonnes of used materials brought in by *Alte Zachen* (Davis & Garb, 2018).

A second major collection source consists of assorted e-waste from institutional organizations, typically collected by larger closed transport trucks with a carrying capacity of seven to twelve tonnes owned by large scrapyards from the West Line villages. These institutional sources range from small computer repair shops to some of the largest companies and recycling facilities in the country, as well as municipalities and other government offices. Regular arrangements and single shipments typically consist of hundreds of single items (computers, screens, refrigerators, entire inventories) when equipment is upgraded at an institutional scale. E-waste from these sources is collected in larger, more uniform lots by Palestinian scrap dealers who have developed standing informal arrangements with key individuals in, or connected with, the relevant institutions. This form of collection is more suitable for large scrapyards, which have a need for and can purchase a continuous supply of more expensive but more valuable e-waste instead of the much cheaper supplies of general scrap from residential collections.

The third collection source, which offers the highest quantities of e-waste, is through a secondary tier of recycling materials already collected from Israeli scrapyards. These yards purchase raw metals and e-waste directly from the public, often by poor and marginalized members of society who pick through neighborhood rubbish bins. Since they collect from many smaller sources, these scrapyards are more reliable destinations for Palestinian buyers than sporadic loads from single large organizations. However, while such pure e-waste and scrap metal loads from Israeli scrapyards are reliable and far more valuable per tonne, they do not come as cheaply as residential or institutional collections. The seller in the scrapyard mediates between source and resale and derives a livelihood from this trade, so they remove their own profit margin and are far more aware of the value of materials. Thus, each load will be negotiated between the Israeli seller and Palestinian buyer, with much room for variation based on the quality and composition of heterogeneous loads and the volatility of metal prices. This latter category of e-waste can only be supported by larger Palestinian scrapyards with the savvy and ready capital to make such purchases, the capacity to transport these large quantities to their facility, and the resources and staff to sort and dismantle it rapidly. Between the institutional sources and Israeli scrapyards, Palestinian collectors from the West Line villages can obtain about 51,800–58,800 tonnes of e-waste annually (Davis & Garb, 2018).

A fourth and marginal source is collection of domestically generated e-waste and scrap metal within the West Bank itself, typically in the larger cities of Hebron, Bethlehem, and Ramallah. Domestic collection is done by Palestinians from the West Line villages without the capital or connections to hire a driver to collect e-waste in Israel. Because of the lower household income in the West Bank and the higher number of people motivated to collect and extract value from domestic e-waste, the amounts and kinds of e-waste available tend to generate lower profits. Typically, this domestic collection is only profitable when the market prices for copper and aluminum are high enough to offset the costs of collection. Thus, domestic Palestinian e-waste is a negligible fraction of the overall trade compared to Israeli e-waste.

Since the bulk of e-waste arriving in the West Line villages derives from Israel, flows are affected by fluctuations in the geopolitical situation and access regime between Israel and Palestine, and, increasingly, to the effects of Israel's regulation of e-waste and the cross-border transport of scrap. Despite these constraints, and beneath the surface of declarations of separation and border closure, the passage of e-waste remains remarkably viable for those motivated to take the risks. Several factors contribute to this porousness of the Israel–Palestine border: the presence of a large population of Israelis living in the West Bank settlements, who pass frequently from one side to the other; the continued availability of gaps in the border regime and the Israeli authorities' pragmatic attitude to merely economic (as opposed to security-threatening) use of these; and the Palestinian Authority's lack of presence at and control over border crossings, which leaves these regulated by only the Israeli side, with its focused agendas. Occasionally the Israeli Authorities, and rarely the Palestinian Environmental Quality Authority (EQA), will issue a press release and photos of an apprehended truck carrying scrap or construction waste providing evidence of their supposed commitment to preventing the cross-border movement of hazardous waste, but these are dwarfed by the scores that flow without hindrance.

The regulation of the passage between Israel and the West Bank was not designed for the regulation of goods. Trucks (including e-waste loads) crossing from Israel to the West Bank have historically been subject to only lax registration or inspection. Once a car has entered the West Bank, there is no tracking system to determine whether it is entering a settlement or driving to a Palestinian village. The entry of trucks to Israel often requires security inspection, but there is less regulation of cargos. Since all entering drivers (including those with e-waste loads) are Israeli residents and drive trucks with Israeli license plates, border guards can assume they are driving directly to or from an Israeli settlement, which is not differentiated from Israel in commercial or tax terms. Theoretically, e-waste loads could be destined for or coming from a settlement, technically never leaving the "country."

This patchwork of shifting circumstances allows well-informed scrap collectors to evade border inspections through various strategies. Israeli citizens (including Jewish settlers) can be hired to transport e-waste loads for the critical few kilometers of border crossing – most exports of motherboards, for example, was historically done via settlers, who would purchase and transport the materials out as their own, not invoking any tax or security checks. Crossings can be done at night when checkpoints are understaffed and less stringent or at the crossings operated by the army, such as the "tunnels" checkpoint, located south of Jerusalem, which is less stringent than those outsourced to private contractors (see Figure 2.1).[4] Some collectors shared with us that they send an advance scout to pass through before a load and call back after passage to signal whether the short-staffed environmental police – the David Unit that explicitly monitors environmental crimes at the border discussed in Chapter 3 – or other hazards are present. And, as discussed in the previous chapter, some collectors may have standing arrangements with the Civil Administration to serve as an informant in return for hassle-free e-waste import at the border. Once the loaded collection trucks enter the West Bank and travel to their destination in the West Line villages, the dispersal and sale of e-waste and used materials begins.

[4] Since the "tunnels" checkpoint is one of the busiest border crossings, with thousands of vehicles passing daily, there are strong incentives for border guards to avoid congestion, only pulling over the occasional suspicious-looking vehicle. Smaller border crossings will scrutinize each vehicle entering Israel, and conduct full airport-style security checks of all non-Israeli passengers and their vehicles.

Figure 2.1 A Palestinian collector approaches the Israel–West Bank border crossing at the "tunnels" checkpoint south of Jerusalem with a load of e-waste and scrap metal.

2.2 E-Waste Processing in the West Line Villages

E-waste arriving in the West Line villages enters a complex network, where it is sorted and then directed to resale, refurbishment, or dismantling. Actors range from several large enterprises with million-dollar annual turnovers at one extreme, through medium-sized scrapyards operating with several workers, to backyard operations in which families handle some subniche of the waste stream, and, at the other extreme of the spectrum, children who pick through the ashy remains of burnt cables, gathering a kilogram or two of copper a week. The larger e-waste-specific loads collected from Israeli scrapyards and institutional sources are typically delivered directly to the scores of large scrapyards. Smaller or mixed loads collected by *Alte Zachen* (dominated by furniture and other household items, rather than e-waste) are auctioned off in the open-air scrap market operating every evening in the village of Beit Awwa. Though a fraction of all materials collected by both scrapyards and *Alte Zachen*, especially the larger electrical

Figure 2.2 An overview of the Beit Awwa Market, in the early evening, as it is beginning to get busy with buyers and sellers of used Israeli materials. Active since the 1960s, this auction thrives from the difference between the availability of low-priced or discarded items gathered in Israel and the demand for used goods in the competitive Palestinian market.

items, are serviceable enough for repair and/or reuse by Palestinians, most are dismantled for the valuable metals they contain – copper, aluminum, and ferrous materials are the economic backbone of the e-waste industry. The key sites and passage points in this feeding chain are described in more detail below.

The Beit Awwa Market is a chaotic combination of a flea market and auction where *Alte Zachen* trucks bring their daily haul to profit from the differential between the free or low-cost items from residential sources in Israel and the high selling price for used goods in the competitive Palestinian market (see Figure 2.2). Dating back to the 1960s, the market grew such that by the mid-2010s, when we entered the scene, it hosted sixty to seventy trucks selling over 100 tonnes of used materials (including e-waste) daily. This is the largest hub of the generalist scrap trade, bustling with hundreds of buyers each evening, ranging from families looking for inexpensive furniture and appliances, refurbishment businesses looking for electrical and electronic equipment to repair, to smaller scrapyards lacking the finances or access to independently collect large quantities of e-waste in Israel, who purchase unrepairable e-waste and scrap metal. The market has generated spin-off businesses, such as several restaurants and food carts, which benefit from the heavy foot traffic, and a dozen or so auctioneers, who come with their megaphones and will attract a crowd to the truck's offerings for a negotiated price, typically ₪25–50 ($7–14), depending on the size of the load.

The Beit Awwa Market feeds a vibrant refurbishment and reuse sector in Beit Awwa, which has been one of the town's main economic foundations for the last thirty years, considerably before the proliferation of the e-waste dismantling industry. Dispersed along the Beit Awwa Market's main road, over 100 shops have established businesses repairing and reselling furniture, appliances, automobiles, computers, and other electronic devices. Refurbishers typically specialize in one or two appliance types (e.g., computer/TV screens, washing and dryer machines, or electric exercise treadmills), fronted by customer-friendly displays of their products, often shrink-wrapped to be indistinguishable from products in new condition. Though only a part of the overall e-waste industry, this refurbishment sector nevertheless constitutes an important contribution to the village economy while extending the productive life of used electronics. While most refurbished electronics are sold in Beit Awwa itself, several shops have arrangements to sell larger loads of computers or other electronic equipment to electronics shops in Hebron, Bethlehem, and Ramallah. For example, a Palestinian trader who purchases 100 slightly outmoded but still functional computers or screens when a large Israeli business or public institution upgrades their equipment will clean and resell these to Palestinian schools, businesses, or individuals. Items found to be unrepairable in these facilities are dismantled to sell their raw materials.

The items deriving from the market are only a small portion of the overall e-waste processing sector, which extends to a range of facilities pervading these villages. Large and medium-sized scrapyards line many roads, and smaller yards and backyard family operations dot the landscape. The approximately sixty large dismantling yards in the West Line are fed by larger trucks, hauling seven to twelve tonnes of e-waste from Israel, which collectively import 50,000–60,000 tonnes of e-waste annually. Most of these scrapyards line the Wadi Risha

Figure 2.3 Overview shot showing a section of the Wadi Risha area where scrapyards and the emerging e-waste industry juxtapose family homes as well as olive trees and agricultural fields.

area, an ad hoc industrial zone between Deir Samet and Idhna (see Figure 2.3). These scrapyards are typically situated on a plot of one to three dunams (approximately four dunams to the acre), equipped with heavy machinery to move and compact scrap metal, strip cables, or shred plastics. They employ specialized staff such as accountants, transport truck drivers, heavy machinery operators, and dedicated collectors with entry permission to Israel. These facilities can employ upwards of twenty workers and process more than ten tonnes of e-waste and scrap metal daily.

Most of these larger facilities have evolved from small family operations through a virtuous cycle of scale-up. There are several key defining thresholds in this growth: achieving sufficient purchasing power and consistency to build relations with Israeli scrapyards that can provide large regular loads; purchase of a larger truck that offers economy of scale; the ability to bypass the local scrap metal traders and sell directly to the national-scale metal dealers in Jerusalem (described below); and the ability to weather the volatility of metal prices, to buy and sell at advantageous times. An incremental growth trajectory is essential for informal businesses, as they cannot raise capital for a large initial outlay based on a business plan and projected profits. Instead, they must depend on their ongoing sales to purchase new inputs and grow, and in this way, they increase their capacity incrementally. Though the benefits of owning a large scrapyard are compelling, the costs are also considerable. The typical start-up finances would be around ₪100,000 ($28,700), which covers rent, walls to secure the facility perimeter, an office, employees, and scrap metal inputs for the first few weeks. Larger facilities incur further operating costs such as paying twenty-four-hour security services to guard valuable e-waste, as well as fuel and maintenance for heavy machinery.

The more numerous smaller dismantling facilities are typically family-run operations and process around a tonne of e-waste and scrap metal per day, sourced from the larger Palestinian cities, the Beit Awwa Market, or large scrapyards from the West Line villages. This kind of facility is harder to identify because it is located inside homes or enclosed house lots and operates intermittently. Because of this, the total number is difficult to know with confidence, and our estimate is over 100 such household dismantlers. These are generally operated by families that have few means to earn an income in other ways, and have the format of a family business, a culturally resonant model within Palestinian culture (Kawasmi & White, 2010), allowing the benefits of unregulated self-employment with fewer of its risks. The entry costs for this kind of operation are low: E-waste dismantling occurs in an existing residential location, requiring only basic tools (hammer, axe, screwdriver, and, sometimes, power tools and grinders), vehicle access to transport scrap in and extracted metals and waste out, and enough capital to purchase the first load of scrap. This can cost as little as ₪400 ($115) per tonne and can often be taken on loan from larger scrapyards who will buy back extracted metals. Once revenue is consistently flowing into the business, more e-waste can be purchased, setting up a pathway to slow incremental expansion of scale and profit.

Regardless of the scrapyard's size, the extraction of valuable components and disposal of the remainder are done in a context with little regulation, only basic infrastructure, and low incentives for incorporating environmental considerations into the process. The procedure is similar in most yards, regardless of size. E-waste and scrap metal loads are piled in the center of the yard, and some initial sorting is done to identify functional machines and parts. Then, axes, hammers, and grinders are used to dismantle and segment metal components from the scrap pile (see Figure 2.4). Those directly attached to insulation are

Figure 2.4 A Palestinian dismantler uses an axe to separate the valuable copper, circuit board, and other metals from a cathode ray tube (CRT) TV. Rudimentary tools such as axes, hammers, and grinders are commonly used in scrapyards as a cheap and rapid way to separate valuable metals from non-valuable components.

burnt in open fires around the villages, and occasionally in the scrapyard itself, and the metals retrieved after the insulation has been burned or melted away.

This sorting process creates large amounts of non-valuable scrap, particularly plastics (computer and screen casings, refrigerator hulls, etc.). The larger yards can afford a driver, a truck, and the tipping fee to dump this waste at the municipal waste transfer station in the neighboring town of Tarqumiya, from where it will be taken to a large formal regional landfill. Smaller scrapyards, however, cannot afford such treatment and will either dump their waste on land they own or on vacant land they do not own, bury it for a small fee in a homemade landfill owned by residents in the West Line, or simply burn it.

These smaller dismantlers do not have the economies of scale to accumulate enough metal to justify independently transporting it to the national-scale metal traders and exporters in East Jerusalem, and, instead, rely on another class of scrapyards, which purchase and aggregate extracted metals from smaller local dismantlers. Once sufficient quantities have accumulated, loads of specific types of metals (e.g., copper, iron, aluminum, stainless steel) are transported and sold to one of the fifteen to twenty such aggregation facilities in the region, especially in Idhna. The trucks are weighed before and after unloading the metals, and the larger trader will offer cash for the weight differential. The offer depends on the cash the trader has on hand, the load's purity (e.g., copper obtained by burning is considerably less valuable than that extracted by stripping), and the current price of metal on the international commodity exchange using the industry standard, JP Morgan (a financial services company that freely displays current metal prices).

These larger local aggregators provide some basic processing of metals (e.g., compressing aluminum into cubes for efficient shipping or sorting grades of metal based on purity) and can negotiate better prices with Israeli middlemen because of the large quantities they deal with. Some of them increase their client base by advancing cash upfront to small scrapyards allowing them to purchase e-waste to dismantle under an agreement that the resulting metal will be sold exclusively to the aggregating trader. This phase of the metal trading business is regarded as the most profitable link for Palestinians in the scrap business, but it requires a great deal of savvy, strong nerves, trust relations, and large cash supplies continually on hand.

Beyond these core activities, as e-waste recycling grew to become the dominant economic activity in the West Line villages, supplemental businesses emerged offering services to e-waste dismantlers. These include homemade landfills, a trucking industry, cable burners, and even children who scavenge metal scraps throughout the villages. Most of the informal commercial landfills are located on the outskirts of the West Line villages in Area C, where Israel prohibits construction, reducing the value of land. Some landowners charge to accept and mix unsellable waste with farm topsoil.[5] The trucking industry

[5] One dominant informal dump site was located in Deir Samet and operated between September 2011 and June 2012. This 2,000 m^2 site was originally opened as a means to economically raise the ground level of the land, making it flush with the main road. Scrapyards paid ₪50–150 ($14–43) per load (depending on the quantity and type of waste) to dump on this property. For every truck load of scrap dumped in his property, two trucks full of soil would be added and mixed with scrap materials (costing ₪150/$43 each). After ten

Figure 2.5 A Palestinian teenager burns cables to extract copper from plastic insulation.

includes not only the vehicles for "external" scrap collection in Israel, described earlier, but also those that haul valuable and discarded materials locally, among villages.

Cable burning to extract the valuable copper from the insulating plastic is the most visible and damaging e-waste recycling practice (see Figure 2.5). Several of the larger scrapyards with high quantities of cables prefer not to burn them directly themselves, due to the personal health and environmental concerns and the risk of getting caught by Israeli or Palestinian authorities. Instead, they will sell insulated cables to "specialists," who burn as a primary income source. While the health risk to these professional burners is large, earnings can be substantial, allowing unemployed teenagers with few employment alternatives a relatively easy opportunity for a decent income. Beneath them in the e-waste feeding chain are the children who seek or wait for active burn sites to scour, brushing through highly toxic ash

months, the owner was able to raise the ground level of a 2,000 m^2 valley by twelve meters, making it flush with the main road. When the final layers of soil are levelled out, the owner planned to plant and cultivate olive trees. This site was just one in a sequence of such dump sites, each operating for a period before reaching their capacity to be replaced by another elsewhere.

in search of remaining copper scraps. This high exposure "copper grabbing" is most common with young teenagers, who will often wait along roads leading to main burn sites, offering their labor toward burning cables (setting the fire and spreading burning cables around with a long iron rod to speed up the process), in exchange for exclusive access to the copper scraps left behind.

2.3 Export Pathways

Ultimately, most extracted metals and all circuit boards are transported out of the West Bank and then shipped abroad through the mediation of Israel-based intermediaries. This mediation is necessary both to ease the transfer to Israel, and because direct export of goods by Palestinian companies is an almost impossibly grueling bureaucratic process that can be navigated by only the largest and most well-established metal traders. This is one of several ways in which Israel's sustained occupation of Palestinian areas has created lucrative niches for Israeli businesses to profit from the limitation of mobility and access of their Palestinian counterparts. The main outputs are motherboards and various metals, with somewhat different mechanisms.

The Palestinian e-waste industry is distinct in not removing precious metals locally, but exporting these intact. The hazardous metal leaching process common in hubs in India (Rochat et al., 2008), China (Puckett et al, 2002), and elsewhere depends on acids, whose availability is sharply restricted in the West Bank – as potential precursors in the production of improvised explosive devices (IEDs),[6] they are in the restricted "dual use" substances list. Instead, circuit boards from scrapyards go to one of three Palestinian collectors, who accumulate large quantities for sale to one of two Israeli settlers, who then transport shipments across the border to Israeli traders for export to refineries in Europe. Though the Israeli settlers act as buyers, their main role is to transport the circuit boards from an Israeli settlement under the guise of "Israeli products," thereby bypassing the importation taxes and security hassles that Palestinian goods would incur (Davis, 2013). The circuit boards are then delivered to the actual buyer, an Israeli

[6] One collector tried to simply burn circuit boards with the hope that the remainders from the ashes would be gold, silver, and other precious metals. Over 800 kg of circuit boards were burnt and consequently left valueless, representing a financial loss of approximately ₪25,000 ($7,175).

trader selling to European refineries. This circuit board transport operation provided the two settlers with a mildly risky but lucrative income source for many years. Their position as one of the only means for export, with little competition, allowed them undue leverage on the terms and pricing of sales. For example, sellers complain that they are forced to lump higher- and lower-value types of motherboards together into a single lower-priced batch.

Copper, aluminum, stainless steel, and other extracted metals are sold to one of two Arab-Israeli traders based in East Jerusalem, which allows them a border-straddling dual status: They accept materials from Palestinians and also own a sister business in Israel that can export metals from their Palestinian to Israeli business without making a "sale," thus avoiding taxation. Interviews with these traders revealed that a Jewish-Israeli partner is still required to ship the metals overseas since Israeli laws restrict the quantity of scrap metal Arab-Israelis can export. This arrangement with the East Jerusalem consolidators affords West Line traders a hassle-free way to send their materials onward on a cash basis, so that they can immediately purchase more e-waste to process. Selling directly to buyers in Israel or internationally would yield a higher price for their metals but would have high bureaucratic overheads, and they would not receive full payment for several weeks, and many are nervous that they may not receive full payment at all.[7]

These East Jerusalem traders have a similar aggregation business model as the smaller metal traders in Idhna that sell to them, but at a national scale, with a clientele stretching across the West Bank and the

[7] The necessity of Israeli facilitation to provide mobility, security, and tax evasion services for the sale of extracted metals back to Israel was not always required. Prior to 2005, informal Palestinian scrapyards would commonly sell metals directly to scrap metal traders in Israel through fictitious companies registered under the name of a deceased Israeli. In doing so, the Palestinian scrapyard could collect a 17 percent value-added tax (VAT) on metal sales, which is supposed to be reported and sent to the Israeli tax authorities, but would not be disclosed, but rather pocketed as extra profit instead. It was common for Palestinian scrapyards to have several fictitious company names at any given time to evade detection from the Israeli tax authority. This practice was so common that an unofficial office in Hebron connected Israelis and Palestinians to purchase fake company names, selling for around ₪10,000 ($2,870) per company name. To the dismay of Palestinian scrapyards, this underground market ended in 2005, when the Israeli tax authority amended the tax on scrap metal, which eliminated this loophole (Israel Tax Authority, 2005).

ability to export metals internationally. Each stage in this destruction network exploits the restrictions of the level beneath it, but prices metals at a slightly higher fixed percentage of the fluctuating global commodity prices. Together, these two traders receive daily shipments of fifteen to thirty tonnes of copper alone, in addition to stainless steel, aluminum, and other metals, all of which they purchase on a cash basis for around 3 percent less than the full JPMorgan commodity price at which they sell. At such large volumes, even a small percentage difference between buying and selling prices can net a substantial profit. For example, fifteen tonnes of copper purchased for $6,000 per tonne and then sold for $6,180 would give an East Jerusalem trader $2,700 profit daily for this metal alone.

In this way, the West Line e-waste economy reflects – in fact, is dependent on – the unequal power relations inherent to Israel's prolonged occupation of the West Bank. As with other economic spheres (agriculture, quarrying, telecom, etc.), the sustained de-development of the Palestinian economy, the inability of the Palestinian Authority to control the borders of the West Bank, and its weakened enforcement powers within much of its area all create an uneven playing field for Palestinian trade. The loss is not only economic, as these asymmetries also facilitate and sustain the most harmful practices within the e-waste sector, and limit the possibilities for Palestinian actors to change these.

2.4 Occupational Hazards

Like other e-waste hubs across the Global South, the significant economic benefits of the West Line e-waste industry carry grave environmental and health costs. We estimated that in the West Line villages alone, 57–64,000 tonnes of Israeli e-waste (not including scrap metal and other used materials) were imported annually, directly supporting over 380 businesses, and more than 1,000 full-time jobs, generating over $28 million in gross value added (GVA) (Davis & Garb, 2018).[8] E-waste has become the West Line's dominant economic sector, whose

[8] The estimated number of jobs and GVA generated by this industry are based on a study conducted in 2015 when the industry was at a low-point due to lowering international metal prices and increased pressure from Israeli border guards to clamp down on this industry. The West Line's historical contribution to jobs and GVA before 2015 was around 40 percent larger than this.

benefits ripple out beyond those employed directly in it to almost all the 45,000 people living in the West Line villages.

This industry is sizeable enough to be a major influence on the Israeli e-waste sector, and even to register within the Palestinian export economy at a national scale. For many years, the West Line villages processed somewhere between 42 and 54 percent of the estimated 120,000–138,000 tonnes of e-waste generated annually in Israel (Balde et al., 2017; StEP, 2016b). While it is difficult to assess a sector mostly unaccounted for in official trade statistics, and muddied by the kinds of cross-registration practices we have just described, the trade in used materials registers a significant part of the trade between Israel and the Palestinian areas. Palestinian purchases of e-waste and other used materials represented 1–1.3 percent by dollar value of Israel's exports to Palestine, and in the return direction, extracted metals represented just over 10 percent of Palestinian export income to Israel (PCBS, 2014a). The GVA from the West Line e-waste industry is equivalent to 0.29 percent of the West Bank's gross domestic product (GDP) (PCBS, 2014b) and 2 percent of the GVA of all industrial activities in the West Bank (PCBS, 2013). These figures are striking for an industrial activity that is almost entirely informal, with no tax revenue accruing to the Palestinian Authority.

These economic benefits, however, come at a cost to public health and the environment due to the regular use of hazardous dismantling methods. And, similar to their economic benefits, while these costs begin within the dismantling facilities themselves, they extend far beyond, to the sites where burning and dumping take place and to the homes of workers who carry the toxins with them. After decades of intensive unregulated operation, these substances and their effects saturate the West Line, affecting the lives of almost all its inhabitants. Here, we focus on the direct occupational hazards, with the environmental impacts of burning discussed in detail in a later chapter.

Modern electric and electronic items contain a stunning array of complex constituents, whose risks when improperly processed have become increasingly clear over time. For this reason, some of the worst offenders have been banned from manufacture, but are still prevalent in the older items entering the e-waste stream. Safe processing of e-waste is a challenge even in North American facilities under strong occupational health and safety standards, where workers have been found to be unsafely exposed to these chemicals (Kuo et al., 2019; Stubbings et al.,

2019). The unregulated recycling methods common in e-waste hubs drastically increase these health and environmental dangers.

Numerous studies of similar e-waste hubs across the Global South show that the burning of e-waste plastics, CRTs, batteries, and mercury-containing components – whether for extraction of metal or disposal of non-valuable scrap – can release a range of toxins. (For informative reviews, see Grant et al., 2013; Sepúlveda et al., 2010; Song & Li, 2014a.) Key among these are heavy metals and various organic pollutants: the remnants and breakdown products of flame retardants (especially polybrominated diphenyl ethers (PBDEs)), polychlorinated biphenyls (PCBs) used in dielectric fluids, lubricants and coolants in older capacitors, transformers and generators, and dioxin-like compounds and polycyclic aromatic hydrocarbons (PAHs) generated during the typically incomplete combustion of plastics. Open burning of insulated cables, auto parts, and other electronics causes incomplete combustion of the insulation containing PBDEs and releases into the air a dangerous mixture of polychlorinated dibenzo-p-dioxins and dibenzofurans (PCDDs/PCDFs) and PAHs. Ash residues from e-waste burning contain heavy metals including lead, cadmium, and hexavalent chromium, among many others, which can be carried on wind-borne particles (Brigden et al., 2005; Gullet et al., 2007; Hicks et al., 2005). "Smash-and-grab" dismantling methods to extract valuable metals from electronic items are also of concern. For example, lead-filled CRT computer monitors are regularly shattered to collect the low-quality circuit board and a small amount of copper located in the neck of the CRT (see Figure 2.5). The valueless glass portions, containing lead oxide, are either left in the scrapyard or discarded in informal landfills where lead can leach into the soil (Tsydenova & Bengtsson, 2011). In addition to the toxic materials, the dismantling of the air conditioners and refrigerators that comprise a substantial portion of the waste stream releases coolants such as Freon, deemed "super-greenhouse effect" gases and strictly controlled under the Kyoto Protocol (see Figure 2.6). Even "clean" e-waste processing, such as refurbishment and reuse, has been reported to cause the inhalation of fumes during electrical soldering operations (Prakash et al., 2010). Many of these pollutants remain in the environment for an extended period, disperse away from the point of release through various mechanisms, and can bioaccumulate in the body and environment, creating secondary long-term exposure paths long after the primary e-waste activities have ceased.

Figure 2.6 Freon is released from an air conditioner after a dismantler cuts it open with a grinder in a scrapyard in Idhna.

When we began work in the West Line villages almost a decade ago, our focus was primarily on the e-waste feeding chain and opportunities for redirection, and only later were we able to obtain the resources for studies now underway to systematically examine the presence, dispersal, and impacts of contaminants stemming from the industry. But from the start, even before directly measured confirmation of their presence, anecdotal evidence of the consequences of e-waste contamination was pressed upon us by almost everyone we spoke with in the community. Below, we highlight a selection of the most alarming stories of e-waste contamination within the agricultural sector that emerged during interviews and observations with farmers and residents in the West Line villages during our fieldwork.

Conversations with olive grove owners revealed drastically reduced quantities of olive oil: Some pointed to trees near burn sites, sometimes more than 200 meters away, which produced smaller and drier olives, reducing the quantity of olive oil they produced, which others told us was bitter to taste. People picking their olives complained that a day of harvesting left them with hands resembling those of a mechanic – a result of the ash that accumulated on the leaves. Subsequent systematic measurements showed that, indeed, this ash was present in worrying quantities, with elevated levels of lead from airborne transport detectable up to 100 meters away from a burn, and waterborne transport extending a kilometer more in some settings. Farmers with field crops located near burn sites reported stunted growth, increased plant pathology, and changes in the taste, texture, and color of their crops. These impacts were later confirmed through controlled greenhouse tests, in which vegetables grown in varying concentrations of burn site ash showed drastically lowered germination rates, lowered biomass, and heavy metal uptake into plant tissues.

Shepherds reported physical symptoms of exposed sheep, including a runny nose mixed with blood, eye abnormalities, excess mucus in the mouth, increasing numbers of physical deformations in newborn sheep, and instances of miscarriages, sometimes up to 50 percent.[9] Pathology occurred not only in sheep and goats grazing in contaminated areas but also in their offspring. Even suckling lambs that remained far from burn sites were exposed because of their mothers' ashy teats. Dairy farmers reported that smoke from e-waste burning reduced the quantity of milk their cows produced by up to 80 percent and increased respiratory disease. Chicken farmers claimed that the quality of eggs declined with the increase in e-waste burning, with shells so thin they often crack before being collected, and we saw for ourselves the prevalence of "dwarf eggs" with little or no yolk.[10] Regional beekeepers we interviewed reported that they had abandoned their attempts to sustain hives in the West Line villages, as smoke from

[9] Sheep not only suffer from the air-borne contaminants from e-waste burning (i.e., inhalation and contamination of grazing areas), but also from direct physical contact with burn sites. When a herd encounters a roadside burn site, they are often left with no option but to walk over it, kicking up ash that is then inhaled. Sometimes sheep get CRT glass stuck in their hooves, and some shepherds report that sheep will chew on plastic remainders from e-waste sites, which leads to inflamed stomachs within a day and often death shortly after. Many sheep drink from a natural spring which is heavily contaminated from surface flows during rain events. Sheep that graze openly are much more likely to get sick or die than sheep that eat manufactured food. Local butchers can even tell whether a sheep has grazed or been fed commercial feed based on the discoloration of meat.

[10] One significant burn event, reported as the largest in the history of Beit Awwa, clued in chicken farmers to link the smoke emitted from e-waste burning directly to the poor health of their hens. The morning after the burn event, the two largest chicken farmers in the village lost all their hens: one had 4,000 and the other 5,000.

e-waste burning had eliminated hives so often that they gave up trying to recolonize them. And, in addition to the loss of these food sources, the products that do survive are likely to contain significant amounts of contaminants affecting public health.

Conversations with hundreds of residents from the West Line villages revealed that one or more of the harmful consequences resulting from e-waste open burning are understood to some degree by nearly every adult resident, and people commonly link the emergence of e-waste burning to an excess of health complaints. In the early years of our work, some downplayed the risks or ascribed excess disease to the prevalence of cousin marriage within the small number of extended families that populate the villages, but this stance became rarer over time. The prominent health complaints reported by the local doctors and community members we spoke with are respiratory problems, kidney malfunctioning, spontaneous miscarriages, congenital disabilities, and cancer (especially lung, colon, and leukemia), whose high prevalence is a source of compelling concern in the villages. For example, when asked about the issue, the mother of one of our translators sat down and quickly produced a detailed list, from memory, of people in her neighborhood diagnosed with cancer in the past five years: a total of thirty-seven people within a two-kilometer radius of her house, with a subsequent more systematic survey indicating a somewhat larger number, with indication of spatial clustering in areas with the most sustained exposures. Our own analysis revealed a striking spatial association between the location and intensity of burn sites throughout the Hebron Governorate and a previously discovered cluster of elevated cancer incidence, with a 4.10 risk ratio for childhood lymphoma ($p = 0.0023$) (Bailony et al., 2011; Davis & Garb, 2018). An increased prevalence of cancer in the West Line villages aligns with a growing body of public health studies that associate cancer etiology with high atmospheric concentrations of heavy metals and organics near e-waste dismantling hubs in China (Huang et al., 2016; Wang et al., 2012; Zhang et al., 2012; Zhao et al., 2009; Zheng et al., 2016). Other studies found that high levels of lead and chromium in the floor dust of e-waste dismantling facilities exposed workers to elevated cancer risks (Ka et al., 2014; Zheng et al., 2013).

Workplace hazards are highest in the dismantling and burning phases, where exposure is through inhalation of fine particles from saws and smoke, and dermal contact. Workers have little protective equipment, if any, evidenced by the scars and missing appendages frequently seen in scrapyard workers. While e-waste workers are heavily exposed to harmful substances and injury risks, of course, there are wider circles of impact. Families are exposed to materials carried home on clothes and shoes, inhabitants of buildings are exposed to the plumes from burning, and our measurements show diffuse low-level exposures to roadside soils along main thoroughfares and dust near e-waste processing activities and in rooftop water from rain.

The international literature indicates that children will be especially vulnerable to lead and cadmium poisoning as they absorb more of these heavy metals from the environment than adults, and their hand-to-mouth behavior further increases their risk (Baghurst et al., 1993). Public health studies in e-waste hubs in Guiyu, China, and Agbogbloshie, Ghana, found that children from these towns have higher blood lead levels (BLLs) and blood cadmium levels (BCLs) than children in reference towns where e-waste dismantling is not prevalent (Huo et al., 2007; Zheng et al., 2008). Even more concerning was that BLLs and BCLs increased with age, suggesting that exposure to lead and cadmium remained chronic. Exposure to these heavy metals is known to cause a wide range of disorders such as damage to the blood and nervous system, neurodevelopment deficits (IQ levels, memory, language, motor skills, and behavior), damage to the kidneys and reproductive systems, as well as increased occurrences of cardiovascular and gastrointestinal diseases (Brigden et al., 2005; Chen et al., 2011; Haefliger et al., 2009). After almost two decades of cumulative exposure, e-waste is posing a growing burden of health risk in the West Line villages. With every passing year, the tension of toxic livelihoods – the dilemma of critical economic benefit at the expense of crippling health and environmental costs – becomes more pointed.

2.5 Concluding Remarks

Our tracing of the West Line e-waste destruction network is emblematic of global transboundary e-waste trade and their destinations in hubs throughout the Global South, but occurring in remarkable spatial contiguity and economic coupling between e-waste source and destination. Similarly, the rebound of improperly disposed e-waste toxins that many developed countries are beginning to experience through poisoned foodstuffs or through the accumulation of e-waste materials in the global commons of the ocean and migrating wildlife is far more direct in the Israeli-Palestinian case compared to the relatively dilute, distant, and delayed impacts of the global trade. Israel's "out of the way" disposal is to Palestinian burn sites that are directly upstream in an immediate and literal sense. This proximity of flow and impact, on the one hand, and the social and political-economic disjunction and asymmetry of these two societies, on the other hand, make this a particularly vivid and intensified microcosm of the dynamics and consequences of Global North–Global South e-waste circulation globally, and of the pathways to reform them.

As we show in the coming chapters, the lens of hubs as global rubbish bins flattens a far more complex situation, masking the vibrant, vertically integrated and intricately interlocked specialized activities and support systems in these places, and their active outreach. Our reframing is not simply an intellectual rehabilitation of local agency but key for quite different and better policies, which revolve around different key actors and solutions. Conversely, the flattening account of hubs on the receiving end of rapacious global dumping sets up bad policies whose effects are mixed, if not negative. As we discuss in more detail in the following three chapters, the kind of sweeping policy response that would save hubs through uniformly restricting Global North–to–Global South e-waste trade via the Basel Convention, for example, ignores the histories, livelihoods, expertise, agency, and development desires of the e-waste-dependent communities they would strive to protect. Thus, they are likely to perpetuate rather than relieve deprivation and health and environmental consequences in e-waste hubs.

In Chapter 3, we detail such conceptions and their policy correlates, and discuss why we take issue with them – both at the regional (West Line) and global levels. Then, in the two chapters that follow, we describe a different and hub-centered alternative. Hub-centered policies build on and upgrade the vitality and hard-won strengths of these places, rather than "saving" them through elimination. Here, too, we describe both the regional case (the focus of Chapter 4) and suggest how this can and should be scaled globally (Chapter 5). Hub-based framings and solutions must, necessarily, emerge from and be responsive to local needs and particular geographies, and policies must emerge from these, building on the strengths and desires of hub inhabitants who know these best. For this reason, Chapter 4 is also methodological, describing how we worked with hub communities to elicit and map desired e-waste futures.

3 *Crude Portrayals, Crude Proposals*

The border between Israel and the West Bank is often portrayed through imagery of towering walls, surveillance, and rigorous security, which belies the daily passage of hundreds of tonnes of e-waste through a single checkpoint or the extensive burning of cables and waste materials immediately adjacent to the notorious border wall without restraint for years. Somehow, the massive cross-border supply chain for the West Line e-waste industry evaded sustained attention and interventions from both Israeli and Palestinian national government authorities for well over a decade. Only fragments of its scope and seriousness surfaced, and responses were ad hoc isolated encounters. This was the case when we began our research, and early on, we decided to initially remain below the radar of public attention.

Several factors shaped our impulse to tread carefully and quietly as we studied and began to work with the local community on this delicate topic. In the early phase, we were still uncertain of the apparent large extent of this industry's economic and environmental consequences. Given that our findings might imply a raised risk of cancer and other health consequences, serious cross-boundary hydrological contamination, including to the important Western Mountain Aquifer, and stigmatization and devaluation of agricultural lands and products, we wanted to have a reliable picture of the phenomenon to avoid false and irreparably damaging claims. We wanted to know what we were talking about, first.

We were also concerned that a narrow and exclusive focus on environmental contamination would lead to policies aimed at eradicating this industry, which would strip the primary source of income from the West Line villages, driving up unemployment while leaving an enduring toxic legacy. Such an undesirable, hostile outcome has been a common government response in other e-waste hubs (Chi et al., 2011; Davis & Garb, 2015; Little, 2018; Reddy, 2015; Wang et al., 2021) and seemed likely to occur in the West Bank based on the recent, strikingly similar, conflict surrounding the charcoal industry in the Jenin Governorate (located along the Green Line at the northern tip of the West Bank). Similar to e-waste burning in the West Line villages, the pollution-heavy charcoal industry emitted plumes of hazardous smoke that crossed over the separation wall, raising serious concerns from Israeli neighbors. With increasing citizen pressure to eliminate this pollution source, the Israeli Civil Administration declared the charcoal industry a form of "chemical warfare," which was leveraged as justification for the Israeli military to enter Area B (a geopolitical area in the West Bank under Palestinian civil control) and demolish scores of family-owned charcoal operations (Surkes, 2016). It was easy to imagine a similar fate for the West Line villages, as sentiments of "terrorism-by-toxins" have been expressed in Israeli media sites (Israel, 2019). We felt that solutions and change must balance contamination and livelihood, local needs, and broader ecological ones, but we did not yet know what this would look like.

Our choice to resist publicity, postponing or restricting opportunities for media coverage or scholarly publications, and to compartmentalize some aspects of what we heard at a local level from briefings we were asked to give to government forums, was aligned with practices endorsed by others engaging in community-based collaborative research projects. Certain sensitive research topics can merit a form of ethnographic refusal: a stance in which researchers and research participants decide not to disclose particular information that could do harm (Simpson, 2007; Tuck & Yang, 2014; Zahara, 2016).[1] Yet, since we were regularly crossing the border, dealing with border guards, poking around the West Line villages, and meeting with influential stakeholders, we may have inadvertently drawn excess attention to the West Line e-waste hub – an inevitable tension within community-based action research that exposes illegal processes, thereby attracting calls for state attention to them.

Despite our cautious approach, the West Line e-waste hub gained increasing public attention around 2012–2013. Several nongovernmental organizations (NGOs) became aware of the environmental contamination attributed to Israel "dumping" e-waste in the West

[1] One of the ways this materialized was by embargoing Davis' master's thesis on the topic, which was the first comprehensive account of the industry (Davis, 2013).

Line villages through personal contacts with residents, which led to influential advocacy campaigns targeting Israeli and Palestinian authorities. The advocacy message had little consideration of the history, dynamics, and scale of the e-waste trade consisting of a singular focus on the negative environmental and health consequences, and strident calls for both Israeli and Palestinian authorities to eliminate the transfer of e-waste from Israel to the West Bank. While this superficially seemed to be a justifiable nationalistic, environmental, and ethical response, their framings of the issue and its policy outcome were at odds with the realities of the industry and aspirations of the villages closest to the phenomena.

These crude characterizations of the West Line villages echo in microcosm (and were implicitly derivative of) international portrayals of other e-waste hubs in the Global South. Global portrayals of e-waste hubs have resulted in policy responses that ban e-waste trade from the Global North to the Global South and establish new formal recycling systems that would reduce inputs to these hubs. Ironically, these do not remedy the underlying economic conditions that fostered the emergence of e-waste hubs nor their enduring environmental harm. In fact, prevailing conceptions of environmental justice for e-waste hubs have produced policy recommendations that stand to harm the marginalized populations they intend to help.

This chapter draws on the West Line case to explore how and why e-waste hubs in the Global South have been portrayed as "digital dumpsites" for the Global North. We begin by analyzing early and foundational exposés of e-waste hubs in Asia and Africa by the Basel Action Network (BAN) that portray these sites as harsh symbols of dysfunctional e-waste management systems. These portrayals were intended to shock the public and compel policymakers to prevent e-waste from ending up in these sites and were widely influential. Viewed from the Global North, their framing is justified as it aims to take responsibility for domestically generated e-waste and reduce culpability for environmental harm in the Global South. Yet, they gave little consideration to the livelihoods and established economies of these victimized e-waste hubs. These portrayals are emblematic of what Baviskar (2011) terms "bourgeois environmentalism," where the predicament of the working-class and marginalized populations is abandoned in favor of creating an "ordered environment." As we show here, these global portrayals of e-waste flows and destination sites and their policy derivatives resonate with those that emerged in the Israel–Palestine case, and suffer similar limitations.

3.1 The "Digital Dump" Narrative

Initial, and still dominant, portrayals of the transboundary movement of discarded electronics and the subsequent emergence of e-waste hubs in the Global South gained traction through a series of influential reports self-published by the BAN – a global watchdog nonprofit organization with a mandate to end toxic trade. Their 2002 report entitled "*Exporting Harm: The High Tech Trashing of Asia*" followed by a 2005 report "*Digital Dump: Exporting Re-use and Abuse to Africa*" (Puckett et al., 2002, 2005) exposed e-waste export from America and industrialized rich countries to a handful of "digital dumpsites" in Guiyu (China), New Delhi (India), Karachi (Pakistan), and Lagos (Nigeria). These reports and BAN's ongoing watchdogging of e-waste exports became the authoritative account of the global e-waste crisis, widely cited in the news media, academic literature, and even policy briefings informing national and international legislation (Lepawsky, 2018).[2]

The two BAN reports consist of a somewhat caricatured portrayal of the massive transfer of toxic refuse from overconsumption in the Global North to "digital dumpsites" in the Global South, fixated on the grave human–environment consequences. The "digital dump" narrative is structured as follows. E-waste is a pressing problem caused by the increased generation of electronics combined with increasingly shortened lifespans attributed to planned and perceived obsolescence and an absence of domestic e-waste management. Insufficient enforcement of international e-waste trade laws, low labor costs, and lax environmental regulations in the Global South have encouraged much of the Global North's e-waste to arrive in hubs in Asia and Africa, which serve as galvanizing icons of the dark side of globalization and the digital age. Within these "digital dumpsites," rudimentary dismantling methods, such as the open-air burning of cables and acid stripping of circuit boards, cause severe environmental and public health

2 Our discussion and critique of Extended Producer Responsibility (EPR) policies and their implementation in Israel, and throughout the Global South, extends a line of argumentation that was initially published in *Environment and Planning E: Nature and Space* (Davis & Garb, 2019).

damages. Here, marginalized populations are forced to choose between poverty and poison and bear the environmental burden of the world's overconsumption of toxic electronics (the reports in this genre typically illustrate these with heart-breaking photos of children in toxic waste sites).

The harsh reality of these e-waste sites is then used to reimagine a more ethical and environmentally just way to manage e-waste in the developed world, centered around two main recommendations, primarily targeting American policymakers. The first is to ban hazardous e-waste exports from developed to developing countries by ratifying the Ban Amendment to the Basel Convention on the Control of Transboundary Movements of Hazardous Wastes and Their Disposal. The second recommendation is to establish EPR policies to make producers responsible for collecting and recycling their electronics and electrical equipment at their end-of-life, which is believed to incentivize the design of electronics to be free of toxic components, and easily upgradable, repairable, and recyclable.

The "digital dump" narrative and advocacy work of BAN did much to raise global awareness of irresponsible e-waste management and its damaging consequences in the Global South. Their framing of the e-waste problem, however, did not emerge from or in dialog with the Global South. Their grounding and target audience were in the Global North, who were urged to take responsibility for managing domestically generated e-waste and take a "do no harm" approach with respect to Global North–Global South e-waste trade, which translated into a call to delegalize it. The message intended for policymakers and citizens in the Global North has successfully inspired and mobilized governments, multilateral organizations, NGOs, private businesses, and academics to expend resources and energy to solve the e-waste problem. The increased global attention and uptick in e-waste policy adoption following BAN's reports and advocacy efforts were a major achievement. Yet, these were structured by the "digital dump" narrative and had unconsidered consequences.

BAN's exposés of e-waste destination sites in the early 2000s – and especially Guiyu, Lagos, and Agbogbloshie – established these as iconic global e-dumpsites, an image reinforced in the portrayals by some of the most respected media organizations over the last decade. Journalists label Guiyu as the "electronic wastebasket of the world" (Watson, 2013), describe "children piling e-waste into giant mountains" in Ghana (Acaroglu, 2013), and refer to Agbogbloshie as "a hotspot for digital dumping" (Siegle, 2017), an "electronic wasteland" (Stein, 2013), or, more commonly, "Sodom and Gomorrah" (Adjei, 2014; Kaplan, 2015; McElvaney, 2014). These imageries are crafted to stir the global conscience, spur individual consumers to recycle properly so that their personal electronic devices do not end up on these sites, and inform sustainable national e-waste stewardship policies and practices (EPR systems, e-waste trade bans). Yet, these accounts and policies are based on thin, often anecdotal evidence and selective engagement by and with environmental activists, NGOs, and journalists that center the easily gained story of poor working conditions and excessive pollution arising from refuse dumped by rich countries. (The authors of "*Exporting Harm*" spent just three days in Guiyu and even less time in Karachi and New Delhi.)

While the "digital dump" narrative captures some of the structural forces shaping e-waste flows, it simplifies the predicament of marginalized populations economically dependent on e-waste recycling. Were the inquiry to engage these people, it would, in all likelihood, learn about their efforts to source and purchase waste as a valued resource, of their vibrant entrepreneurial and productive activities, such as reuse, repair, and materials recovery or the industry support services. Ironically, the dumping narrative disconnects e-waste flows from the broader historical and postcolonial terrain of injustices that led specific regions to become e-waste destination sites in the first place (Akese & Little, 2018; Davis et al., 2018), and from calls for restorative action to assure the social and economic upgrading of these places. It also disconnects images of e-waste's health and environmental consequences from the responsibilities and opportunities for remediation of land and treatment of the public health burden.

The register in BAN's reports and subsequent media exposure is primarily moral, in which the sender of e-waste from the Global North imaginatively takes the position of its recipient. "A free trade in hazardous waste leaves the poorer peoples of the world with an untenable choice between poverty and poison – a choice that nobody should have to make" (Puckett, 2002: 2). In arguing for a ban on e-waste trade, BAN positions itself as the moral arbiter of the global allocation of harms and benefits. "[E]ven if it were somehow possible for China, India or Pakistan to possess state-of-the-art technologies, and the resources and infrastructure to ensure that such technologies work optimally, the export of all of the world's e-waste to Asia would still be an unjust, inappropriate export of pollution to a particular region of the world simply because it is poorer" (Puckett, 2002: 3). Here, BAN attempts to close the moral door to e-waste exports by denying the developing world the capacity to recycle e-waste and incorrectly assumes that all the world's e-waste is shipped to regions in Asia (Lepawsky & McNabb, 2010). One might wonder whether BAN would claim it is unjust that the world's end-of-life circuit boards are shipped to a handful of state-of-the-art refineries in developed countries (i.e., Aurubis, Boliden, Dowa, LS Nikko, Umicore, and Xstrata), where they are profitably recycled with environmentally sound operations. Somehow, BAN manages to speak on behalf of the best interests of marginalized populations in these "digital dumpsites" without much dialog with them.

In what follows, we offer a brief overview of the two central policy instruments proposed by these reports – EPR policies and trade bans – highlighting how they stand to harm, rather than help, the living conditions of e-waste hubs in the Global South.

3.1.1 The Ban Amendment to the Basel Convention

The primary advocacy message of BAN's reports, and one of their organization's driving missions, is to enforce a global adoption of the Ban Amendment to the Basel Convention. The Basel Convention itself is an international treaty designed to reduce hazardous waste trade between nations and prevent transfer of hazardous waste from developed to less developed countries. The Basel Convention was initially conceived in response to the increase in hazardous waste trade in the 1970s and 1980s. As developed nations began tightening environmental laws, the disposal costs of hazardous waste grew significantly. At the same time, the transboundary movement of hazardous waste became more accessible through the globalization of shipping. Developing countries desperate for foreign currency became a convenient and economical means to dispose of hazardous waste from the developed world.

The mediatization of several high-profile toxic dumping cases gave urgency to regulate hazardous waste trade. The year 1986 marked the beginning of the infamous Khian Sea incident where 14,000 tonnes of incinerator ash from Philadelphia, which was no longer accepted for disposal in New Jersey, was loaded on a cargo ship searching for a place to dispose of it. Of this, 4,000 tonnes were dumped in Haiti, labeled as "topsoil fertilizer," before the Haitian government became informed of the waste's origin and forced them away. After months of unsuccessful efforts to unload the ash in a handful of countries, it finally "disappeared" somewhere between the Atlantic and Indian Ocean in 1988. Shortly after, another high-profile case of toxic dumping occurred in the small fishing village of Koko, Nigeria, where 8,000 barrels of hazardous waste, disguised as building materials, were shipped from Italy and improperly stored on farmland resulting in the contamination of the surrounding environment and population. A Nigerian farmer received $100 in monthly rent as compensation. These toxic dumping incidents led Greenpeace to coin the term "Toxic Colonialism" to describe industrial waste dumping from the Global North to Global South and were the driving force leading to the Basel Convention (Pratt, 2010).

However, shortly after the Basel Convention's adoption in 1989, environmental activists denounced it as an instrument that legitimized and regulated hazardous waste trade rather than outright prohibiting it. The main point of contention was the "prior informed consent" clause, which permits the transfer of hazardous waste from developed to developing countries so long as the sending and recipient countries grant consent (Clapp, 2001). In response, a coalition of countries and environmental NGOs passed the Ban Amendment to the Basel Convention in 1994. The Basel Ban Amendment effectively bans the export of all hazardous wastes (including e-waste) from developed nations (i.e., Annex VII countries: Organization for Economic Cooperation and Development (OECD), the European Community (EC), and Liechtenstein) to the developing world (i.e., Non-Annex VII countries). At the time of writing this, the Ban Amendment has yet to enter into legal force as three-quarters of the member states to the Basel Convention have yet to ratify it. However, the Ban Amendment is honored by most Basel Parties.

A criticism of the Ban Amendment and other blanket policies in general is that they tend to imagine everything being the same everywhere, overlooking the unique and fluctuating dynamics of specific places and situations. For example, Lepawsky (2015) demonstrates the bimodal imaginaries embedded in the Basel Convention, which groups all countries into one of two categories based on inclusion in the OECD. This creates some strange outliers, such as Singapore and Hong Kong being classified as "poor" countries. To some, these bans have been labeled discriminatory, denying entrepreneurs in the developing world access to valuable resources based on their ethnicity (see www.WR3A.com). The incinerator ash and industrial waste cases that gave urgency to the Basel Convention have a negative value and are fundamentally different from e-waste, which can often be repaired and resold on the secondhand market, or the raw materials recovered by recycling. Moreover, the Basel Ban does little to stem the tide of e-waste that arrives at the now-notorious "digital dump-sites," since much of the e-waste at these sites either is domestically generated or arrives from the Global South (Breivik et al., 2014; Grant & Oteng-Ababio, 2012; Lepawsky & McNabb, 2010). Because the Ban Amendment offers a blunt policy instrument that strictly bans Global North–Global South e-waste trade, it has struggled to gain the same momentum and acceptance as the Basel Convention and has been opposed by many from e-waste destination sites in the developing world who profit from the import of e-waste from the developed world.

3.1.2 Extended Producer Responsibility (EPR)

About a decade after the Basel Convention entered into force, national e-waste management became an increasingly prevalent domain of environmental regulations as countries developed and implemented specific legislation for the collection and disposal of e-waste (Oliveria et al., 2012). In Europe, where national e-waste management first emerged, e-waste legislation has been framed by two directives passed by the European Union (EU) in 2003. The Waste Electrical and Electronic Equipment (WEEE) Directive requires member states to encourage the design and production of electrical and electronic equipment that recognizes and facilitates their dismantling, recovery, reuse, and recycling (Sinha-Khetriwal, 2006). The Restriction of Hazardous Substances (RoHS) Directive legislates against the inclusion of certain hazardous substances in developing electrical and electronic equipment, reducing adverse consequences of poor end-of-life management. Environmentally friendly substances must replace hazardous substances such as lead, mercury, cadmium, hexavalent cadmium, polybrominated biphenyls (PBBs), and polybrominated diphenyl ethers

(PBDEs). Ideally, these two directives function in harmony to address e-waste at the beginning and end of the pipeline.

Under the impetus of these two e-waste directives, many countries have adopted EPR-based e-waste management policies, which are considered by both the OECD and the EU as the most promising means to address the increasing generation of e-waste and pollution from improper treatment (Langrova, 2002). Under EPR legislation, the responsibility of e-waste shifts upstream (both financially and physically) to producers and manufacturers. This is believed to provide incentives to "design out" environmental problems by keeping the product's entire lifecycle in mind, which will reduce resource use and improve product recyclability (OECD, 2001). Moreover, EPR helps relieve the swelling financial burden of e-waste management from municipalities by shifting the responsibilities and recycling costs upstream to producers and manufacturers.

While EPR programs are tailored to the implementing country's local contexts, Manomaivibool (2009) identified three critical elements in any e-waste EPR program. First, a Producer Recycling Organization (PRO) must be established to ensure e-waste is recycled in an environmentally and socially acceptable manner. PROs will spur the creation of authorized treatment facilities (ATFs) to collect, repair, and recycle e-waste in a controlled manner. Second, producers must internalize their products' end-of-life consequences. Ideally, producers should be responsible for collecting and treating their products, providing an incentive to design electronics that are cost-effective to recycle. Lastly, monitoring and reporting mechanisms need to be embedded within an EPR program to establish targets and verification markers that provide insights to evaluate the system effectively.

Policies based on EPR principles are often advertised as a tool that will address e-waste at both the beginning and the end of the pipeline. In practice, however, virtually all EPR schemes focus exclusively on the latter (Plambeck & Wang, 2009). Several academic and policy reviews have criticized EPR policies for having weak or nonexistent environmental design incentives (Atasu & Subramanian, 2012; Gui et al., 2013; Rotter et al., 2011; Sachs, 2006). Lepawsky (2012) also points out that EPR policies place the financial burden of collecting and recycling e-waste on consumer-citizens, discouraging producers from incorporating design for the environment in electronics. Thus, the primary outcome of EPR has been end-of-the-pipe solutions to create incentives and channels for waste collection and proper treatment. The global push for EPR has resulted in seventy-eight countries adopting national e-waste policy, legislation, or regulation that have collectively recycled 9.3 million tonnes of e-waste annually, accounting for 17.4 percent of global e-waste generated (Forti et al., 2020).

As European countries became leaders in e-waste management, they began sharing their rationales, approaches, and systems with other countries. One avenue of this knowledge transfer was through Basel Convention Regional and Coordinating Centres, a network of fourteen centers predominately located in the Global South.[3] These centers focus on training and technology transfer regarding the management of hazardous waste, including e-waste, and host workshops and conferences to raise awareness of the e-waste problem and share best practices. Other avenues of knowledge transfer have been through international research hubs such as Solving the E-waste Problem (StEP) and the United Nations University, multilateral organizations such as the World Health Organization and International Telecommunication Union, and nonprofit organizations, such as the International Solid Waste Association.

However, the realities, challenges, and solutions of European e-waste management do not neatly extend to other contexts, and several studies have discussed the idiosyncrasies of e-waste management in a developing country setting, problematizing the application there of EPR models (Nnorom & Osibanjo, 2008b; Osibanjo & Nnorom, 2007; Sthiannopkao & Wong, 2013). For example, European countries are less likely to import e-waste or host a pervasive informal sector that has historically collected and processed most national e-waste. Most e-waste in Europe is believed to be improperly landfilled or exported to the developing world, thus creating economic opportunities for value capture and the requirement of an organized e-waste management system. E-waste is referred to as an untapped "urban mine," with only $10 billion recovered from the potential $57 billion value of raw materials generated each year (Forti et al., 2020). Thus, the combination of minimal competition from the informal sector and the opportunity for value capture has enabled Europe to achieve a

[3] These regional and coordinating centers are located in: El Salvador (closed down since May 2017), Trinidad and Tobago, Argentina, Uruguay, Russia, Slovakia, Senegal, Nigeria, Egypt, South Africa, Iran, Indonesia, China, and Samoa.

collection and recycling rate of about 42.5 percent of total e-waste generated in 2019 (Forti et al., 2020).

In contrast, countries in the Global South tend to have low labor costs, weak environmental regulations and inadequate enforcement, and a market for secondhand electronics, and import e-waste. Thus, there is likely to be a robust informal e-waste sector that collects e-waste locally and processes large quantities of imported e-waste, all on an economic commercial basis rather than as the result of regulatory mandates and subsidies. Even now, most of the world's e-waste, particularly in the Global South, still moves through largely informal channels. Thus, while e-waste is generated globally and diffusely, the processing of e-waste primarily takes place at the end of informal collection networks, which tend to agglomerate in spatially defined hubs throughout the Global South (Davis et al., 2019). Despite this stark difference in the circumstances and functioning of informal sectors and hubs in the Global North and Global South, over forty non-OECD countries with economic and governance conditions that would suggest the presence of an active informal sector have now established national e-waste legislation, policy, or regulations based on European models. Not surprisingly, these non-OECD countries have far lower collection rates than OECD countries, less than ten percent of total e-waste generated compared to four times this in OECD nations (Forti et al., 2020).

Thus, the emergence of formal systems that use policies to mandate the collection and recycling of e-waste in competition with informal ones that already do so robustly in practice, has given rise to contradictions and tensions. For policymakers and decision-makers, informal systems were initially neglected, or considered marginal, or portrayed as victims of improper recycling practices that EPR would supersede. Policies did not consider the volumes flowing through informal systems, the highly competitive prices they could offer, and the advantages of unregulated and tax-free operation, even if these undermine formal systems and collection quotas (Akenji et al. 2011; Ciocoiu, 2012; Davis, 2020; Kojima et al., 2009; Manomaivibool & Vassanadumrongdee, 2011). In addition, the formal systems tended to disadvantage vulnerable populations by shifting resources and legitimacy from the least to the most capable social strata, often with no clear environmental and health gains. Formal systems target high-value/low-hazard e-waste from lucrative commercial sources, marginalize reuse and repair economies, and tend to relegate the low-value/high-hazard e-waste from residential sources that are more challenging to collect to the informal sector (Kama, 2015; Reddy, 2016). As the scale of the informal sector became apparent, it was met with gradations and admixtures of antagonism (against the pirate industry), a patronizing stance (dismay at the fate of the "dumped upon" and attempts to protect them from this), or aspiration for some kind of partnership (best articulated in the "best-of-2-worlds" strategy, which we discuss in more detail in Chapter 6) (Davis & Garb, 2015; Wang et al., 2012). While often well-intentioned, such partnership efforts usually emerged from the formal waste-generating side and, at best, offered informal businesses a pathway to assist, transition toward, and be incorporated within a newly established formal sector.

In this sense, both the Ban Amendment to the Basel Convention and EPR policies are forward-looking approaches that redirect e-waste away from "digital dumpsites" to formal and regulated e-waste channels. In doing so, however, they largely ignore the past, ongoing, and future realities of e-waste hubs and the livelihoods of marginalized populations that depend on e-waste recycling. While they prominently reference harmful working conditions and environmental pollution of e-waste hubs for legitimacy, these policies do not address these conditions nor boost the ability of affected communities to address them, whether by upgrading capacities or transferring to other forms of livelihood. In the end, they risk leaving them with a triple hazard of reduced inputs, a legacy of pollution and injury, and a new regulatory landscape whose definitions, legislation, and enforcement practices marginalize and even criminalize their practices.

3.2 Portrayals of the West Line E-Waste Hub

In many ways, the global "digital dump" narrative resonates with how Israeli and Palestinian actors have portrayed the West Line e-waste hub once environmental NGOs and media sources first publicized the issue. This framing led both sides to initiate policies that would eradicate the unwanted e-waste sector, though informed by different motives. From the Israeli perspective, the illicit transfer and hazardous dismantling of e-waste were viewed as a source of cross-border environmental contamination and a "leakage" that undermined the national EPR system's viability. From the official Palestinian perspective, the inability to control their borders enabled the structural conditions for unwanted Israeli e-waste to be smuggled into their territory. The West Line villages became emblematic of an undesirable outcome for both sides and fodder for broader nationalistic narratives.

A more detailed understanding of the historical and political drivers that led to the West Line e-waste hub's emergence and its economic importance might suggest an alternative narrative. For example, this could focus on Palestinian entrepreneurs, who with considerable ingenuity and against incredible odds, developed an e-waste economy into one of the few forms of available livelihood in an oppressed and constrained economy – indeed, into one of Palestine's larger but least visible export sectors. Given that this industry obtains over 90 percent of its inputs from Israel, and directly or indirectly supports most households in the area, the stoppage of e-waste flows to the West Line would be tantamount to eliminating the industry and likely to be fiercely opposed locally. Yet, these polarized portrayals of the West Line villages – by Palestinian authorities as hapless victims of toxic colonialism or by Israelis as agents of reckless avarice at best, or chemical warfare at worst – both converged into calls for similar policy responses. These policies prioritized reducing environmental harm over the livelihoods of the marginalized populations benefiting from – but also suffering most directly – irresponsible e-waste management. In what follows, we trace how Israeli and Palestinian government officials became exposed to the West Line e-waste industry, which shaped their initial characterizations of it and the resultant policy responses.

3.2.1 Israeli Portrayals and Proposals

The Israeli authorities became increasingly aware of and pressured to respond to the West Line e-waste hub through two isolated interactions: Israeli complaints of e-waste burning registered with the Civil Administration and the Ministry of Environmental Protection's (MoEP) implementation of an EPR-based e-waste recycling system.

With around ten tonnes of e-waste burned daily in the West Line villages, which creates massive pillars of black smoke, sometimes stretching kilometers on end, it was only a matter of time until the neighbors started asking questions. Indeed, Israeli residents from the small settlements of Adora, Tellem, and Negohot – along with the villages of Shekef and Eliav on the Israeli side of the Green Line – witnessed frequent occurrences of thick black smoke plumes that led to noxious smells since the emergence of the e-waste industry and, increasingly, raised concerns for their own health. Initial speculation was that the smoke resulted from solid waste disposal burns due to inadequate waste management or, possibly, politically charged protest burns of tires, which are common in times of high political tension. Without a presence in the villages nor direct communication with Palestinian municipalities in the area, the causes and materials burnt remained unknown and served as a growing source of frustration. This led to frequent complaints to the Civil Administration demanding swift intervention, and the Civil Administration responded with forceful efforts to eliminate e-waste burning. These tactics began by deploying military jeeps during active burn events, which delivered the occasional arrest of Palestinians caught burning, and at times violent confrontations, such as one publicized case that led to the death of a Palestinian teenager (Levy & Levac, 2018). However, the frequency of burn events and their short duration made it resource-demanding and logistically challenging for the Israeli military to apprehend burners, which rendered this approach largely ineffective.

Considering this, the Civil Administration shifted toward a preventative approach. In 2014 they implemented a ban on certain materials believed to be destined for burning (i.e., cables, tires, and cathode ray tube (CRT) TVs) from entering the West Bank. The David Unit monitored this ban, which comprised a team of six to twelve border guards commissioned initially to prevent the illegal dumping of used construction materials in the West Bank. Trucks found with contraband items (typically larger transport trucks collecting e-waste from institutional sources) were confiscated for about a month and fined around ₪7,000 ($1,850). However, rumors circulated of special arrangements to circumvent such penalties for those willing to cooperate as informants. In practical terms, this ban only slowed the flow of e-waste materials into the West Bank. Palestinian collectors learned how to evade the David Unit, as discussed in Chapter 2. Overall, the Civil Administration's initial response to burning in the West Line villages was combative, a game of cat and mouse with the David Unit and military personnel chasing e-waste smugglers and burners. While a nuisance to those working in the West Line e-waste industry, those practices did little to reduce e-waste burning.

At the same time, increasing attention toward domestic e-waste management was surfacing within Israel proper, influenced by European framings for e-waste policy. As part of preparing for

Israel's acceptance into the OECD, several environmental recommendations were made to break the link between economic growth and environmental deterioration. In response, the MoEP passed an EPR-based e-waste law in June 2012, entitled *Environmental Treatment of Electrical and Electronic Equipment and Batteries Law*. This was drafted and promoted by an Israeli environmental NGO, the Israel Union for Environmental Defense (Adam Teva V'din). This legislation placed responsibility on product producers, manufacturers, and importers to properly recycle product-derived e-waste. And, like EPR schemes in other OECD countries, Israel's e-waste law did not pressure producers and manufacturers to address e-waste management in electronics design and focused exclusively on end-of-the-pipe solutions by establishing regulated channels for e-waste collection and recycling. This joined Israel's prior solid waste management policies, following existing legislation on packages, tires, and bottle recycling. The e-waste law came into effect in 2014, allowing local authorities, retail stores, and manufacturers time to prepare new collection and treatment systems (Adam Teva V'din, 2012).

Israel's e-waste law set in motion a new government-sanctioned economic market and formal e-waste industry as manufacturers and importers must finance the collection, recycling, and treatment of e-waste through one of two certified PROs[4] (i.e., MAI and E-community).[5] Within this system, Israeli e-waste is legally owned by the two certified PROs, who then subcontract e-waste collection and recycling responsibilities to accredited companies. E-waste collection is restricted to three channels: local municipalities, retailers of electronics, and holders of electronics (e.g., banks, companies, etc.). Each of these collection channels is obligated to transfer collected e-waste through one of the PRO's collection systems. The two PROs are required to collect e-waste at a rate of 15 percent of the average total weight of electronic equipment sold by importers making use of their services[6]

[4] In the Israeli e-waste law, Producer Responsibility Organizations (PROs) are referred to as "Accredited Compliance Bodies" (ACBs). However, PROs are the common terminology used in the EU and North America, which, for the sake of consistency, we adopt in this book.

[5] Several other entities have applied to become an accredited PRO, though, to date, no others have received accreditation.

[6] Many importers of electronic and electrical equipment have yet to register with Israeli PROs, making collection targets much lower than the percentage of all Israeli e-waste.

during the previous two years (calculated biannually), to be ratcheted up by 5 percent every year until 2021, when the collection target would reach 50 percent. The law was premised on preventing e-waste from entering landfills, converting e-waste into a national resource, and encouraging the local recycling industry to create new green jobs for Israeli citizens and residents (AdamTeva, 2012; Shammai, 2017). Our interviews with the Israeli MoEP and the Israeli Union for Environmental Defense revealed that both institutions were unaware of the existence – indeed, overwhelming importance – of the Palestinian e-waste sector when developing the Israeli e-waste law. This blindness to the existing informal Israeli-Palestinian e-waste system indicates incomplete groundwork when creating a new regulated e-waste market, or, possibly a degree of unwillingness to surface or grapple with years of cross-border hazardous waste trade or the disruption of the livelihoods of Palestinians dependent on Israeli e-waste.

In the first few years of the e-waste law, Israeli PROs struggled to achieve collection quotas. Through a series of e-waste management public consultation conferences and workshops in Tel Aviv as well as our own private meetings, Israeli PROs became increasingly aware of the substantive Palestinian e-waste sector that was (at least in their mind) "robbing" them of their legally owned e-waste and undermining the functionality of their EPR system. The introduction of a formal e-waste recycling sector as a novel competitor to an experienced informal sector raised various clashes and concerns. First, since producers and importers are not expected to pay consumers to return their discarded electronics to formal collection points, the informal sector has an immediate advantage as they are willing to pay for these products. Second, informal recycling networks currently perform comprehensive and efficient collection services at people's homes. Under the formal system, this service is only offered to residents disposing of large appliances (i.e., fridges, washing machines, etc.) in cities with a population over 100,000. Over the decades, citizens have come to rely on the frequent and convenient door-to-door rounds of informal collectors. Third, consumers often keep e-waste in their homes for an extended period before thinking about properly disposing of it. Contacting a collection facility or taking materials to a collection site requires more initiative and foresight than the current informal situation where collectors regularly solicit homes for their waste and offer immediate cash and convenient removal as an incentive. Finally, retail stores and producers that collect discarded electronics may be tempted to profit by selling their collected e-waste to informal recyclers. Since collection quotas are by weight, with little discrimination of value and toxicity, high-value/low-weight fractions can be sold separately, and weighty low-value and toxic fractions (such as CRT monitors) given to formal channels. Overall, in the eyes of Israeli PROs and the MoEP, the Palestinian sector was a hindrance to meeting collection quotas and undermined the newly established EPR system. This resulted in PROs pressuring the MoEP and Civil Administration to ensure Israeli e-waste, now a national resource, does not leak across the border to the Palestinian sector.

3.2.2 Palestinian Portrayals and Proposals

In contrast to the arm's length connection Israel has to the West Line villages, the Palestinian municipal governments of Idhna, Al Kum, Deir Samet, and Beit Awwa have been painfully aware of the environmental and public health consequences of the e-waste recycling sector dating back to the early 2000s. They had not conveyed the extent and urgency of this issue onward to national levels, however. We expect there may be an intrinsic motivation to downplay the reality that the municipal economy functions as a recycling hub for Israeli rubbish. The transfer of e-waste across the border is quasi-illegal according to both Israeli and Palestinian laws, and the municipal governments would not want to strip their villages of the principal source of income. It is also worth mentioning that several municipal mayors over the past decade have owned some of the largest scrapyards. The municipalities were concerned with their inability to control the controversial open-air burning of e-waste, but the fragmented and often chaotic state of Palestinian governance in the West Bank impeded transferring the situation's urgency to national levels, themselves preoccupied with national and nationalistic struggles. Since solid waste disposal burning and environmentally questionable industries are ubiquitous throughout the West Bank and not subject to serious attention from national governments, it is understandable that the West Line e-waste industry remained off the radar or had a low priority to national Palestinian government offices for well over a decade.

Outside of government channels, however, the West Line villages gained increasing public attention around 2012. A group of environmental activists from Idhna, many of whom are farmers whose crops and livestock had suffered greatly due to the environmental contamination from e-waste burning, began raising awareness of the issue, approaching a range of Palestinian government branches and civil society organizations. Attempts to petition the police, Environmental Quality Authority (EQA), and even the Prime Minister's Office all fell on deaf ears. However, the Applied Research Institute of Jerusalem (ARIJ), one of the largest Palestinian environmental NGOs based in the city of Bethlehem, responded to an invitation to visit the villages. They sent a junior staff member to spend a few days in Idhna investigating e-waste recycling in the area, resulting in a brief self-published report documenting their findings (ARIJ, 2012) .

The report relied on interviews with environmental activists and farmers as the authoritative community voice regarding the e-waste industry. Drawing inspiration from the "digital dump" narrative, they labeled the West Line villages an "electronic graveyard" for Israeli e-waste, providing one more plank in their broader call to end the Occupation. The methodological thinness of the report resulted in outlandish estimates: that the town of Idhna alone (roughly half the population of the West Line villages) received 200–500 tonnes of e-waste daily (equivalent to the upper estimate for all e-waste generated in Israel), generating approximately one billion shekels ($264 million) annually – which would have been double the value of all Palestinian agricultural exports, and on par with the entire Palestinian export sector of all categorized manufactured goods combined. The sources and methods by which these estimates were obtained are not mentioned and are entirely at odds with the same organization's detailed research portrait of the town some years earlier (ARIJ, 2009), which described it as a typical rural village with no indications of an e-waste sector.[7]

The report's framing of the West Line situation prioritized a nationalist agenda over the local population's livelihoods. For example, it

[7] As part of an internationally funded development needs assessment, ARIJ conducted and published an extensive survey of each of the villages in the region. This assessment reported that Idhna (a northern village in the West Line) was a rural village ("30 percent of the households are rearing and keeping domestic animals") whose main public health concern was sewage from the nearby Israeli settlement of Adora. Aside from the mention of twenty junk stores (alongside four hundred grocers and thirty clothing shops), there is no mention of the e-waste industry. Our research shows that the industry was already in its heyday at this point, with extensive environmental impacts already evident.

describes e-waste in the West Line villages as "a result of the unregulated transfer of waste from Israel and illegal Israeli settlements" (ARIJ, 2012: 4). The inclusion of Israeli settlements here is puzzling and serves a specific purpose. Palestinian collection trucks are not permitted into the heavily gated settlements for security reasons; thus, e-waste from Israeli settlements would not be directly collected by Palestinians nor arrive in the West Line villages. Our interviews with e-waste collectors and scrapyard owners strengthened our assumption that they did not source e-waste from settlements. While there could be circuitous extra-legal pathways for settler e-waste to find its ways to the villages, the centering of this component, without supporting evidence, was most likely the use of a familiar trope about injustices of occupation. The report also frames Israel as dumping e-waste in the West Line villages, stating that "Israel has taken advantage of this situation by transferring vast quantities of e-waste into Idhna across the Tarqumiya checkpoint" (p. 11). This depiction downplays the role of Palestinian collectors in facilitating e-waste flows from Israel to the West Bank, very little of which arrives through Tarqumiya, and which Israeli authorities claim to be unable to prevent. ARIJ's report was the first public account of the West Line e-waste hub, and was accessible on their organizational website and later presented to the Palestinian EQA.

In a similar vein, a second Palestinian environmental NGO, the Green Land Society for Health Development (GLSHD) based in the city of Hebron, was aware of the West Line e-waste industry and partnered with an Israeli NGO, the Arab-Jewish Center for Empowerment, Equality, and Cooperation – Negev Institute for Strategies of Peace and Economic Development (AJEEC-NISPED) to develop a project grant proposal aimed at reducing environmental contamination. This partnership secured a three-year €468,000 grant from the European Partnership for Peace program for a project entitled "*Solutions for electronic waste disposal affecting Palestinians and Israelis*" (GLSHD, 2013). The proposal stated ambitious goals to simultaneously advocate for the elimination of e-waste importation from Israel, which represents 90 percent of all inputs to the West Line e-waste industry (Davis & Garb, 2018a), while somehow still facilitating "job creation and financial opportunities for advancement" through clean e-waste recycling (GLHSD, 2013: 2). A few years later, in 2016, GLSHD partnered with the North Hebron Chamber of Commerce and Industry and an Italian humanitarian NGO, CESVI, to secure yet another EU grant. This three-year, €1,250,000 project, entitled "*Promotion of environmentally safe, innovative and economically valuable WEEE (Waste from Electrical and Electronic Equipment) treatment in West Bank*" sought to continue the objectives from their previous grant. This project aimed to provide managerial and business training for scrapyard owners, raise awareness of the environmental and public health harm associated with e-waste recycling in the villages, introduce green technologies, promote improved working conditions, and develop a shared e-waste treatment facility. Similar to the previous grant, this project focused on building a formal e-waste recycling sector in the West Line villages by processing Palestinian e-waste exclusively. Regardless of the local opposition to these projects, or even their economic and technical feasibility, a collective of influential outside NGOs was armed with funding and a self-assigned mandate to raise awareness of the e-waste issue and pursue a specific development trajectory. The effort had only limited engagement with the West Line villages and even fewer results materialized over time, beyond some signs displaying logos and projects' names. Once the funding ended, the facility's operation and the program effort were abandoned, and project staff moved on to other projects elsewhere, leaving only an empty warehouse and resentment locally.

The framing of imported e-waste as driven by Israelis dumping hazardous contaminants in the Palestinian territories resonates with some portions of the West Line population and upward to the sphere of Palestinian national politics in a critical stage of struggle to end Israel's crippling occupation. Declarations and even donor-funded development projects predicated on ending the importation and processing of e-waste in this region (ARIJ, 2012; GLSHD, 2013) emerged readily in this context, and in others internationally, and often can display some post hoc trappings of approval from the local community. Yet, the selective form of community engagement in the West Line villages was embedded in a preestablished narrative imposed by Palestinian NGOs in which the e-waste problem represented a new kind of environmental harm of Israel's prolonged occupation. These NGOs leveraged their recently acquired expertise on the e-waste issue to produce policy briefings and conferences for the Palestinian EQA, focusing on the legal situation of e-waste management in the West Bank, the responsibilities of various Palestinian authorities, and recommendations to reduce environmental and public health harm in the region (ARIJ, 2015).

In turn, the EQA declared its opposition to the import of any e-waste into the West Bank (including items in working condition). This position was likely based on a combination of NGO reports and advocacy stigmatizing the West Line villages as an "electronic graveyard" for Israeli e-waste coupled with Palestine's environmental law that bans the importation of hazardous waste (although e-waste is not explicitly defined as "hazardous").[8] Moreover, Palestine's recent ratification of several international treaties to boost its legitimacy as a nation-state, including the Basel Convention, offered a legislative device for the EQA to hold Israel accountable for permitting e-waste trade to the West Bank as required by the country-to-country mechanism specified in the Basel Convention protocols. However, the EQA's policy stance has been solely declarative with minimal action. Palestine's occupied status, its inability to control its border, limited resources, and fragmented governance all undermine its ability to prevent e-waste from entering the West Bank. Thus, the EQA adopted a somewhat untenable declared strategy of upgrading and regulating e-waste businesses in the West Line villages to process only domestic (i.e. Palestinian-sourced) e-waste while banning e-waste importation from Israel, despite the latter comprising the overwhelming source of the industry's inputs. While these simplified portrayals of the West Line e-waste hub resonated with broader Palestinian political impulses and grievances, they were, in fact, at odds with pragmatic solutions for residents of the West Line villages whose livelihoods and daily exposure to e-waste contaminants demanded urgent and effective action.

3.3 Concluding Remarks

In many ways, the e-waste issue has been conceptualized as a global problem as seen from the Global North. The debates it has generated regarding the ethics of Global North–Global South e-waste trade and predicament of e-waste hubs in the Global South, driven by many

[8] Specifically, the Palestinian Environmental Law defines hazardous waste as "waste generated by the various activities and operations or the ash thereof, which preserve the characteristics of hazardous substance which have no uses, such as atomic waste, medical waste, or refuse emanating from manufacturing of pharmacological products, medicines, organic solvents, dyes, painting, pesticides or any other similar substance." (Palestinian Environmental Law, 1999).

academics, journalists, NGOs, and national governments, are often distant from the environmental and economic consequences as experienced in these hubs. Whether in the West Bank or elsewhere in the Global South, these hubs figure readily as global symbols of irresponsible e-waste management in a way that offers little practical guidance toward an equitable and sustainable approach to remedy their predicaments. The crude portrayal of e-waste hubs as a "digital dump" might be accurate in some ultimate structural sense, and a blanket ban on e-waste imports from the Global North is one possible policy response to this. However, such framings and suggestions did not emerge from a deep strategy for social justice nor evidence-based consideration of policy alternatives, but from environmentalists in the Global North informed by inadequate data and superficial community engagement regarding the nature and extent of e-waste destination sites. This caricatured portrayal resonated with dominant political and environmental currents, at the expense of the more informed and involved stories that would be demanded by closer examination.

Missing from these debates, both internationally and in the Israel–Palestine context, is the perspective and agency of e-waste hub residents, who have borne the environmental burden of processing e-waste, both from domestic and foreign sources, and are most directly affected by e-waste management policies. Ours is a more hub-centered approach to e-waste policies, which posed two essential questions, addressed in the following chapter: (1) How might communities economically dependent on e-waste portray their local economy? (2) What developmental aspirations and policy responses might they advocate? We describe an innovative community engagement method that facilitated a heterogeneous and contentious group of community representatives from the West Line villages to articulate a widely accepted development vision and plan to improve their local economy. The policy stances of national governments and environmental NGOs described in this chapter, diverge starkly from the answers arising from the West Line communities in this manner, which strove to reduce environmental harm while retaining the economic gains from their e-waste industry.

4 Co-creating E-Waste Hub Futures

On a cool October evening in 2012, we attended a community meeting organized by several residents of the West Line villages to discuss the e-waste issue. By this time, we had spent several weeks in the villages learning about the e-waste industry's complexities by observing and interviewing representatives from municipalities and various government agencies, workers and owners of e-waste-dependent businesses, farmers, elders, and just about anyone who would speak with us. One of these interviews was with Layla, the head of the local Women's Society, who stood out as a rare progressive woman in the villages – highly educated, independent, and determined to improve her community's living conditions. The fact she drove a car, a gendered role reserved for men in this region, was a telltale sign she was leading the charge for women's rights. When she learned of our interest in the e-waste industry, she was eager to share her work on the topic as part of a coalition of local civic groups. Along with the Women's Society, members from several civic groups in the villages (e.g., The Higher Education Society, Palestinian Red Crescent Society, Idhna Charitable Society, among others) had been meeting regularly for several years out of concern for the environmental consequences resulting from the importation and burning of Israeli e-waste that, in different ways, overlapped with the organizational missions of each civic group. Their collective mission was to raise awareness of the deleterious effects of e-waste dismantling among actors and institutions able to influence or curb these practices. Their advocacy efforts materialized into various initiatives, such as protests at local municipal and national government offices, small grant proposals to international donors to fund awareness campaigns, and outreach meetings with potential allies, such as the one we attended.

The meeting took place in the Center for Higher Education, a donor-funded public building in Idhna that regularly hosts various health, business, and educational programs. There were six community members present who were hosting representatives from two Palestinian environmental nongovernmental organizations (NGOs) and one journalist, as well as us academics. We sat around a long table in a simple version of a board room, furnished with used items most likely sourced from the Beit Awwa Market. Before the meeting began the standard fare of cola and juice was served in little plastic cups along with an assortment of store-bought pastries and sweets.

As the meeting began, discussion of the e-waste issue revolved around stories of the various environmental and public health tragedies that have plagued their villages. These ranged from the contamination of water, livestock, and agricultural products to increased incidences of rare diseases to an overall decline in the beauty of a once-pristine landscape. Both the Israeli government and those working in the e-waste sector were equally villainized for poisoning their community – the former for enabling the conditions for Israeli e-waste to be "dumped" in the West Line villages, and the latter for polluting their village with little consideration for the well-being of their community. Solutions offered were simple and narrow: stop Israeli e-waste from entering their villages and the harmful e-waste dismantling practices occurring within them.

Curiously, in contrast to the majority of the West Line population, no one at the meeting derived their livelihood from the e-waste industry, either directly or indirectly, nor was the perspective of the economic importance of e-waste given much consideration. Our time thus far in the villages had shown us the devastating environmental consequences of the e-waste industry as well as its economic significance. This meeting, therefore, was the equivalent of a community meeting to discuss the problems and desired futures of regional economies such as coal mining in Wyoming or tanneries in Tamil Nadu, India, without including the perspective of those employed in these industries. This was our introduction to one possible civil society depiction of the e-waste issue, a selective framing drawn from a particular segment of the population that echoed the crude portrayals of Israeli and Palestinian NGOs and government agencies discussed in the previous chapter.

The journalist sniffed out a compelling story of another example of Israeli negligence polluting a victimized rural community, whereas representatives from environmental NGOs saw an opportunity to partner with an accessible and articulate group from the West Line villages to apply for external grant proposals. This framing was readily embraced since the e-waste problem reinforced preestablished narratives of Palestinians suffering from the environmental externalities resulting from the prolonged occupation. The viewpoint biases of a small number of citizens offered a monolithic uncontested perspective as representing the best interests of a heterogeneous and often divided community. But this rendition was one well suited to gain the support of outside allies and resources.

The emergence and circulation of a selective community framing among local citizens for external audiences is not unique to the West Line villages. Similar dynamics are at play, for example, in the infamous e-waste hub in Agbogbloshie, Ghana, which has attracted numerous international environmental NGOs, journalists, photographers, environmental scientists, among other activists. In their brief visits, these actors typically witness intense e-waste burning and its damaging consequences and are informed about the context of the e-waste issue from local environmental organizations and government agencies that have a troubled relationship with the e-waste sector (Akese & Little, 2018; Little, 2016; Minter, 2016). Similar dynamics have occurred in e-waste hubs in Guiyu, China, and Lagos, Nigeria, among other sites (Puckett et al., 2002, 2005). Without "development brokers" that mediate communities to the broader world, journalists and NGOs would likely have less interaction with or awareness of the troubling phenomena of e-waste hubs, which might remain hidden environmental catastrophes. But there are unintended consequences when external allies seek to help these communities without due consideration of the economic importance of e-waste hubs and local livelihoods. In the absence of input from a diverse and representative range of local community members, a symbiosis of a narrower local group with external actors can reinforce a partial worldview and blind spots when they talk to local "communities."

This chapter describes our attempt to break this circuit. Specifically, we wrestle with the complexity, challenges, and importance of representative community engagement and framing in development projects. We explore this in the West Line villages by mapping out some of the social and political divisions in a seemingly homogenous region and our effort to facilitate an unconventional community engagement method that sought consideration of these diverse perspectives. Our community engagement method contributes to the critique and recovery of community representation in participatory development through a novel Delphi-like method that enabled a broadly endorsed development proposal within a heterogeneous and conflicted community. This method offers a new perspective on e-waste hubs in the Global South and problematizes simplistic national and international policies that would exclude them from e-waste economic activities. We show a process that enabled a complex and often divided community to, nonetheless, converge toward a development trajectory based on building capacity and regulating their existing e-waste economy to preserve livelihoods and remedy environmental hazards.

4.1 Toward Representative Community Engagement

The unintended consequences of the kind of shallow or skewed community engagement and framing we saw in the West Line villages have been a topic of much debate in the development studies literature. Since the mid-1980s, the inclusion of communities in externally funded projects through various forms of participatory development has been popularized and even routinized in development practice, arising as a response to structural distortions, not to mention ineffectiveness, of externally imposed forms of top-down development approaches. In practice, these forms of participatory development typically manifest through partnerships with local social groups advocating on behalf of the community and employ visual and tangible consensus-based participatory methods as the basis for identifying needs and developing projects emerging from and beneficial to the entire community (Chambers, 2008; Francis, 2002). The popularity of and demand for participatory methods made it tempting to streamline and package these in a way that could be readily transmitted and replicated. Best practice manuals proliferated, and participation techniques were taught and applied in diverse community settings (Chambers, 2008), often in a way that gained priority over more differentiated and critical reflections of community representation (Leal, 2007).[1]

Increasingly, though, development scholars problematized simplified conceptions of communities, asserting that the time and financial pressures to rapidly engage with local populations have led to

[1] This chapter is based on our previous publication "Participatory shaping of community futures in e-waste processing hubs: complexity, conflict, and stewarded convergence in a Palestinian context" in *Development Policy Review* (Davis & Garb, 2017).

tokenistic, even caricatured, participatory approaches. These undercut community empowerment by partnering with certain social groups in the community, often an elite stratum accessible to development agencies. In doing so, considerable control in the participatory process is retained with an elite minority and the development agency, which channel change in directions that resonate with their interests and perceptions, but under the name of the community (Guijt & Shah, 1998; Michener, 1998; Mosse, 2001). Indeed, the implicit or explicit curation of community members who are consulted and engaged with can depreciate local community realities, exclude marginalized populations, and at worst, be manipulative and constitute a dangerous form of co-optation (Botes & Van Rensburg, 2000; Cleaver, 1999; Eversole, 2003; Funder, 2010). Thus, the banner of participation could allow a kind of collusion between development broker, donor, and interest group in the community to generate "local knowledge" (Mohan & Stokke, 2000) and apparent ownership in support of development interventions.

Some of the trappings of such selective community participation revolve around a set of simplifying assumptions – "myths of community" (Guijt & Shah, 1998) as bounded and homogenous, and of any form of participation as inherently benign and transformative. Community participation is founded on an image of a preexisting and definable community – a distinct bounded entity, defined by administrative, social, and natural boundaries – that development practitioners can tap into and participate with. Yet, such convenient geographic or other demarcations rarely map onto the relevant populations (Sihlongonyane, 2001), as shown by a growing body of literature demonstrating the permeability of borders and the overlapping, shifting, and subjective nature of communities (Cleaver, 1999). An allied simplification is that communities are homogenous in nature, with the role of participatory methods being to tap into their unified voice. Various scholars have analyzed the heterogeneity of communities, fragmented along myriad social, cultural, economic, and political differences (Botes & Van Rensburg, 2000; Hickey, 2008; Parfitt, 2004). Others have shown how misconceptions of personal incentives or the community allow participatory processes to entrench inequalities by reinforcing power in dominant groups through the selective inclusion and exclusion of community groups (Funder, 2010; Mosse, 2001; Platteau, 2004; Rigon, 2014; Sesan, 2014).

The challenges of social divisions within communities are manifest in some of the strategies development practitioners use to recruit community participants, which often favor existing powerful networks and accessible villagers who are vocal, educated, and articulate that become spokespeople for the entire community (Botes & Van Rensburg, 2000; Eversole, 2003). At public community meetings or convened forums, for example, selection can occur through active exclusion, or the more mundane filtering effect of work, time, logistic, or cultural conflicts (Cornwall, 2003; Mayoux, 1995), while even those present may be inhibited in speaking or in what they can say. Community-based organizations (CBOs) that appear as natural partners for sustained participation programs may have leaders that are not democratically elected or may represent the voice of a particular segment of the community, usually the more well-off, with specific interests that can diverge from the needs of marginalized populations (Hayward et al., 2004). The consensus-based group decision-making methods, such as voting or ranking, often used as a participation tool can produce apparently unified problem framings and descriptions of practical needs that mask divergent definitions and interests (Cooke & Kothari, 2001a; Green, 2010). When conducted in open real-time contexts, these can reinforce patterns of dominance and exclusion, yielding an apparent community expression of practical goals without addressing underlying diversity and inequalities from which problems and solutions arise and are voiced (Mayoux, 1995).

Despite numerous critiques and even dismissals of certain kinds of community participation as a masked form of tyranny (Cooke & Kothari, 2001b), many scholars still recognize its value and necessity (Hickey & Mohan, 2004, 2005; Kyamusugulwa, 2013). Approaches have evolved from their initial exuberance, in which inclusion of communities in a rapid and superficial template led to a simplified and idealized conception of them and the participation process, to a more thoughtful consideration of how participatory development processes can account for the unique social and political contexts of each community. Thus, participatory methods strive now to more adequately represent the diverse, often conflictual, cleavages of a target community.

Recent studies try to broaden community participation by taking local circumstances into account when designing community participatory methods (Agarwal, 2001; Brett, 2003; Lange, 2008; Mansuri &

Rao, 2013). For example, Dill's (2009) study on participatory approaches in Tanzania showed how CBOs, which are the traditionally favored participatory partners for development agencies, achieve their legitimacy with external development agencies and governments at the expense of their legitimacy vis-à-vis marginalized citizens from the community they "represent." In such settings, democratically elected local authorities, with downward accountability to the community, offer a more suitable, although often less favored, institution for development agencies to partner with (Ribot, 2003). Kyamusugulwa and Hilhorst's (2015) analysis of elite capture in the Democratic Republic of Congo showed that the dynamic of "elite capture," traditionally considered one of the most significant threats to community-based development, may in some cases serve as an asset. In that case, the dominance of churches and chiefs was shown to ensure the provision of public goods through notions of moral obligation and interpersonal accountability that worked to benefit the community. Thus, uncovering the social structure, power relations, and exclusionary effects of participation within each target community may lead to participatory approaches that seem unconventional yet are more appropriate to the local context.

These community engagement analyses show the importance of a preliminary phase of information gathering and reconnaissance to understand the community's dynamics. This initial investment of time and resources to unpack the social, cultural, economic, and political divisions will sensitize external actors to the potentially harmful effects of development projects informed by and exclusively serving one segment of the community. This contrasts with the tendency toward the rapid participatory tools and reliance on CBOs that have become the favored partners of development agencies to accelerate community engagement and rapidly obtain the necessary information and apparent buy-in required to implement "community-owned" projects. This reconnaissance phase yields a slow participatory process, but sensitizes external actors to more diverse community segments and how development projects may hold differential outcomes within the community.

In what follows, we describe some of the major social and political divisions within the West Line villages and highlight the importance of including these groups in forming a community-articulated e-waste future. We underscore how the unconsidered participatory shaping of the community can lead to a framing of development goals benefiting a small segment of the community, while generating animosity and economic hardship for others.

4.2 Multiple Representations of the West Line Villages

To many, the rural West Line villages would appear to be largely uniform. The population is almost entirely Palestinian and Muslim, and much of the culture, architecture, and traditions seem identical from one household to the next. Yet, these villages are torn and united through divisions – most noticeably pertaining to their family or clan, livelihood, and gender – and one is likely to elicit distinct portrayals of the West Line e-waste industry from local villagers depending on their location within these (Davis & Garb, 2017). Indeed, it was immersion in the West Line villages for nearly three years that revealed some of these deeper social, economic, and political divisions, in which the community is not a preexisting monolithic and static entity to be represented and characterized by any local citizen, but is constituted, shifting, and – even within a snapshot moment – riven along multiple lines of tension and fragmentation. Participatory contact and elicitation with any one segment of the West Line villages would entail selection, power, and exclusion.

Below, we isolate some of the key divisions we found and our attempts to consider and incorporate differences when eliciting an encompassing e-waste vision. Sometimes cleavages run through a single individual, an intersectionality that can create overlapping and interdependent divisions in the community, exacerbating disadvantage, or, alternatively, provide resources and networks for livelihoods and e-waste operations.

The extended family or clan is a key social structure in the West Line villages, especially in the domains of land ownership and marriage (see Figure 4.1). As discussed in Chapter 1, Beit Awwa's population consists of two extended families, the Masalma and Sweity families, geographically divided by a central road. Interfamily marriage is considered dishonorable and condemned on both sides. These divisions are so taken for granted by locals that they are seldom mentioned, though they inform a variety of patterns. An outsider would most likely notice the prevailing force fields along gender or Israeli/Palestinian lines but may not realize that it would be less acceptable for a Sweity man to marry a Masalma wife than a Jewish one.

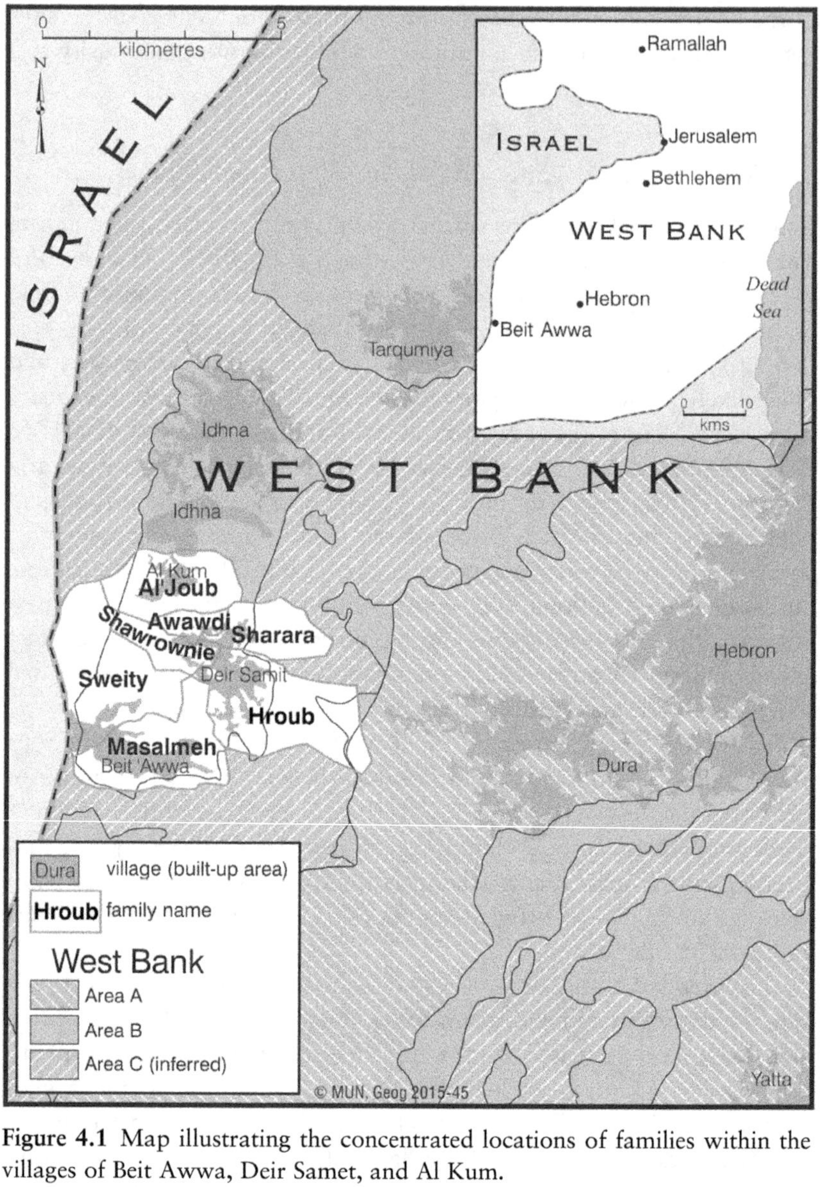

Figure 4.1 Map illustrating the concentrated locations of families within the villages of Beit Awwa, Deir Samet, and Al Kum.

Economic, religious, political, educational, and social orientations are inflected by these clan lines and shape the development goals each articulates. Even municipal councils, which consist of twelve elected community members, are mediated by careful informal agreements to assure at least one member from each family is elected and that key positions will be rotated or offset to accommodate clan divisions. In the past, unrepresented families have protested by refusing to pay taxes or electricity bills when they feel their concerns are overlooked in the municipal council. Family divisions often align with political divisions, particularly between Hamas and Fatah supporters, which represent the two main political parties in Palestine, in conflict since 2006. Customary law offers an alternative forum to the formal justice system in several domains, often revolving around mutual liability for family members' actions and a negotiation process between family heads of conflicting parties to resolve disagreements amicably. As described in Chapter 1, the e-waste industry is family-based, with certain families abstaining from the industry and others dominating niches within it. Thus, knowledge of the family clan that one is speaking with will likely inform their perspective on the e-waste industry.

Differing sources of livelihood present another layer of division, related to, but not strictly tied to, family divisions. Dating back to the Ottoman period and earlier, the West Line economy was based on agriculture and livestock farming, which still generate a significant source of income in the community. The emergence of the e-waste industry, while generating a much-needed livelihood for most of the population, came at the expense of agricultural production, generating new lines of tension, as many farmers complain that e-waste processing has negatively affected the health and profitability of their livestock and crops (Davis, 2013). Farmers (mostly representing an older generation) would like to return to an agriculture-based industry. Those dependent on processing imported e-waste and used materials are more attuned to the economic gains of a scrap-based economy, which they would prefer to continue as is or in some more regulated and formalized version. These livelihood divisions interact with clan structures, with five or six extended families economically supported by imported e-waste and used materials. These tend to be wealthier and less educated, each dominating a niche in the industry (e.g., refurbishment, dismantling, and metal trading). Within each niche, families are located differently with respect to ownership of the means of production (trucks and scrapyards), shaping their stance toward any intervention in this industry.

As in most rural West Bank locations, the roles of men and women in the West Line villages are well defined. Men are responsible for financially supporting the family, while women take care of the children and home. There is often a tension within households economically supported by the e-waste industry; women often have a more ambiguous relation to the industry, at times expressing a desire for family members to shift to another profession regardless of the economic sacrifice, while men want to provide for their family in an economic context in which there are few alternatives to earn a comparable income. Women often complain of the strenuous effort required to remove the grease and black soot that covers their family members' skin and clothes, and tend to be more concerned about the potential health consequences from the introduction of hazardous substances into the home, especially with small children present. However, cultural and social norms limit women's influence and decision-making power, both in formal settings (men dominate municipal and community meetings in numbers and voice) and domestic ones.

With increasing politicization of the West Line e-waste industry, and the emergence of characterizations resonant with only a minority of the region's inhabitants, it became increasingly important to find a way to elicit a community-defined portrayal and management plan that would gain broad support while balancing the development desires of those financially dependent on the industry and those for whom health and environmental concerns were dominant.

This question of navigating to mutually agreed-upon development goals with a broad basis of support in a diverse and sometimes divided and contentious community became quite real around 2014, as we shifted from discovering and researching issues surrounding the West Line e-waste economy which we began in late 2011, toward a more active effort to translate our findings and the desires expressed by the people we encountered into interventions. Our field work, certain incidents, and the literature on these topics sensitized us to the shortcomings of selective participatory methods, such as interviews with a biased advocacy group or seemingly open community meetings narrowly advertised to and attended by individuals opposing the e-waste industry. And beginning in June 2014, we began to explore alternative modalities for community participation, in particular the adaptation of a traditional Delphi technique, described below. This draws inspiration from Kezar and Maxey's (2016) change-oriented Delphi, and demonstrates the apparently contradictory merits of externally selected key informants contributing to project design in a way that ensures representation of core demographics and social divisions in the community. This approach produced the community-articulated vision and strategic pathway toward a desired e-waste future that informed the advocacy efforts with community members, which we describe in Chapters 5–8.

4.3 Facilitating a Shared Vision for E-Waste Futures

Our community engagement involved eight carefully chosen "community representatives"[2] who participated for ten months in a series of communication iterations through a spoke and hub communicative process between them and the facilitators (see Figure 4.2). These communication rounds progressed through four phases: (1) articulate an improved local economy; (2) identify physical and social assets within the community; (3) brainstorm and select specific development projects; and (4) develop and implement selected projects. Each phase involved two sets of interviews, the first collected raw data from a set of open-ended questions, and the second encouraged community representatives to comment on the anonymized responses of the others and rethink their own. Each phase required two to three weeks to complete, which comprised two sets of interviews with all eight community representatives, with each session lasting between one to three hours. This form of off-line and anonymized iteration allowed different, often conflicting, perspectives to have an equal voice and express varying realities in a nonhostile environment, and the range of responses on the table converged on a mutually agreeable position. Moreover, the iterative and long-term engagement with community representatives encouraged a reflective disposition regarding the progress of community-selected projects – identifying expected and unexpected barriers, reassessing original articulated desired development goals, and evaluating the progress and success of each project. As opposed to the traditional Delphi procedure, interactions with community representatives were face-to-face in their homes or places

[2] We chose "community representatives" as the most suitable term for community members participating in the project as they were representing the diverse interests of the community. The term "participants," commonly used in participatory development, denotes a more passive role, whereas the term "community researchers" or "co-inquirers," commonly used in participatory action research (PAR), would indicate a research-centered role.

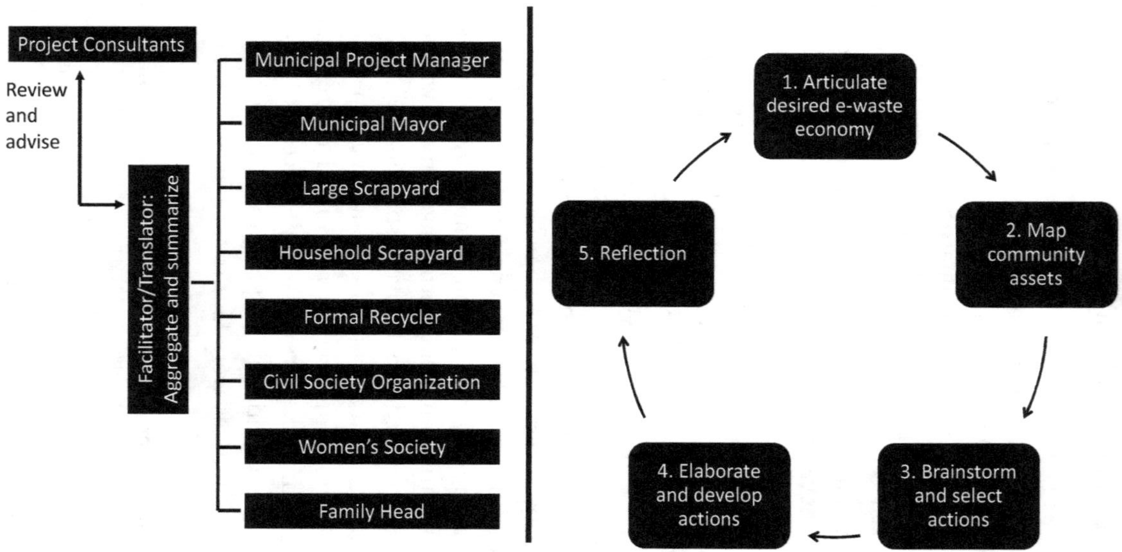

Figure 4.2 A visualization of the adapted "change-oriented" Delphi method. The graphic on the left presents the flow of information for each round of interviews. The graphic on the right presents the iterative progression of community engagement.

of business at times convenient for them to decrease the "cost of participation" (Brett, 2003) since we were unable to financially support their participation.

The number of community representatives was limited to eight to allow the meetings and analysis to advance in a timely manner. Representatives were selected based on over 300 prior interviews we had conducted with diverse stakeholders in the villages, both within and outside of the e-waste industry, coupled with a socio-historic mapping of family and municipal leaders in the West Line villages. Insofar as was possible for a group of only eight participants, we made a concerted effort to represent cross-cutting social and political divisions and divergent interests. Our community representatives included individuals from different families and villages, both women and men, individuals working in different niches of the e-waste industry and outside of the industry, and representation embodying both governmental (municipal authority) and traditional/tribal forms of authority. Beyond our attempt to adequately represent diverse social divisions within the West Line villages, our selection of community representatives also weighed in characteristics of good "community researchers" from the PAR literature, which include a strong knowledge of the community, a passion for improving the community, strong communication skills, ability to think creatively, and deep linkages within the community (Cameron & Gibson, 2005).

Beyond these eight representatives, the process benefitted from the inclusion of a wide circle of "project consultants[3]" whom we were familiar with who could review and advise on project ideas and strategies developed by community representatives. They provided external reality checks and assisted us in identifying external resources. Counter to the belief that successful community development projects are possible solely based on sufficient internal community mobilization, our engagement with these consultants fed important expertise, insight, and perspectives into the iterative community process, forging a more viable and informed identification of opportunities, barriers, and potential community projects. The project consultants explicitly and implicitly drew on information, contacts, and inputs from a range of

[3] Recognizing that the term "consultant" has a range of contested meanings in the development literature, we use this term to denote lay volunteer/pro bono efforts as opposed to a paid or unpaid professional role.

stakeholders in the community and the e-waste industry: scrapyard workers and traders, government officials in the Palestinian Authority, the Civil Administration and the Israeli government, Israeli and Palestinian environmental activists, lawyers, and representatives from international government organizations. Through our mediation and cross-pollination, these project consultants reflected and commented on the impact and feasibility of various emerging project ideas, the political, social, and legal barriers that might complicate their implementation, and ways to improve strategies and proposals. Information from project consultants was included in project reports given to community representatives at the end of the second, third, and fourth phases.

Our role as facilitators was to carefully choose and initiate this process with the community representatives, serve as a catalytic go-between for them and the project consultants, and synthesize and prepare inputs for the rounds of review by community representatives. In fact, the role of the facilitator increasingly became a fused effort of Davis with a local English-Arabic speaker familiar with the area, who worked as a translator and, increasingly over the course of the project, a research assistant. Having a local community member immersed in all aspects of the process assured attention to linguistic and sociocultural nuance, and boosted the transparency of these meetings and the accountability of the external facilitator to the community. This pairing balanced the advantages and disadvantages of each participant alone: The external positioning of foreign nationals and status as academics gave some degree of exemption from some of the local clan and other political force fields, and the obvious naiveté of being an outsider to the villages allowed us (and others) to pose (or simply voice) less obvious ideas or proposals. It also enabled us to reach outward and upward to Palestinian and Israeli decision-makers and stakeholders and, subsequently, potential project funders, who could not be readily accessed by the West Line community.

At the same time as we favored inclusion and transparency, we also worked hard to buffer between the various actors we moved between, and preserve their anonymity. As discussed further below, insofar as it was possible, careful attention was given to bracket the researcher's and translator's personal visions of community development, so that these did not shape nor direct the community representative's responses. In some cases, the proposals that emerged were not ones we would have raised or chosen, but we were gradually persuaded of their merit. Thus, we too partook in the convergence toward a shared set of understandings and proposals.

There is certainly an irony in an intervention that seeks to avoid selective representation and exclusion or silencing of many community members from development discourses and processes, while achieving this through apparently limiting community engagement to a selection of community representatives hand-picked by an outside facilitator. In our defense, we studied and mapped the structures, cleavages, and tensions within the community and, based on these, selected the community representatives, rather than relying on conventional "open" means of community recruitment or partnering with outspoken individuals from the community. We consulted with our community representatives and wider groups in the West Line villages to ensure we were not overlooking key social divisions or perspectives and gained approval from our community representatives and project consultants to proceed using the Delphi method. We were also aware of the participatory sacrifice that comes with inserting ourselves as a principal and necessary point of contact and interpretation in the iterative spoke and hub structure and process. We can only argue that we did this with considerable awareness and trepidation; we employed a series of corrective measures, and the emergent results seemed to validate that the dialog and engagement we hoped for did indeed occur.

Our decision to externally select participants was taken out of a well-grounded sense that some of the more familiar "participatory" selection and engagement mechanisms were likely to produce a group or set of power interests biased toward a specific development trajectory on the one hand, and lead to alienation or even hostility from unrepresented social groups (Cornwall & Jewkes, 1995). Our choice of a Delphi polling method and the location of our role at its hub at the apparent expense of participatory principles was not taken lightly. In the best circumstances, open community meetings have the potential to build a sense of community and strengthen social ties among the local population (Ahmad & Abu Talib, 2015; Purdey et al., 1994) and increase the potential for the community to take control of development projects, rather than putting the onus on the facilitator to organize and drive projects forward (Cornwall & Jewkes, 1995; Ritchie, 2000). Yet, development scholars have increasingly realized that community development projects would typically not occur without the initiative of someone with time, skill, and commitment, which is most often someone from a privileged and educated group (Reason, 1994). The fact that so little is written on the beginnings and endings of community engagement practice attests to the fuzzy conceptions of facilitator and community relations (McArdle, 2008). Our adoption of the change-oriented Delphi method was done only after considerable observation of the community, and especially of the stark, and at times hostile, divisions between families and different development desires in this region, and how these played out in discussion and decision-making fora.

There is obviously a danger that a facilitator will selectively approach and favor individuals, interpret events and statements in a biased way by their own background and interests, or even become manipulative (Botes & Van Rensburg, 2000). Nevertheless, we judged this option to have a greater likelihood of yielding a productive and representative process than a "participation" process centered on partnering with a local social group or hosting open community meetings, that would typically privilege one dominant voice to speak on behalf of the community.

At the end of the day, the apparent productivity of this approach assuaged our anxieties about its participatory purity. Despite complex and often contentious lines of division within the West Line communities, it facilitated an iterative productive articulation of broadly shared representations of the harms and benefits of the e-waste economy and suggestions for concrete projects to grapple with these, which matured into successful proposals for funding support and then pilot projects. In particular, the process helped bypass a long-standing deadlock between the several small community groups and individuals previously polarized by their respective dismissal of the e-waste industry's environmental or health consequences.

4.4 The West Line's Articulated E-Waste Future

As the community representatives moved through the four phases of the Delphi method, iteratively reviewing and commenting on each other's responses, the contours of an inclusive and mutually agreed-upon community-articulated vision emerged. During the first phase, a key issue that emerged immediately was a consideration of scenarios ranging from the continuation of the status quo of an informal economy revolving around processing Israeli e-waste to one that moved toward creating a regulated and clean e-waste economy to one that worked to stop the e-waste trade and shift the economic basis of the villages toward alternative livelihoods. Community representatives then generated a comprehensive list of the advantages and disadvantages of each path. After the first round of questions, considerable differences were apparent between individuals on whether the benefits of imported e-waste outweighed their negative consequences. All community representatives recognized the considerable economic importance of imported e-waste; what varied was the perception of associated environmental and public health risks and the plausibility of reining these in, or of developing alternative livelihoods, especially between those employed in this industry and those outside of it.

As community representatives reviewed each other's responses, there was a clear convergence on two points of shared understanding. The first was that the e-waste industry was too economically significant to this region to shut down, yet it could not continue in its current pollution-heavy form. As one community representative employed outside of the e-waste industry stated, "had it not existed, the scrap industry would be the most beneficial development project that they could propose." The second was that all community representatives believed the e-waste industry could be regulated and improved. Thus, already in its first phase, the anonymized semi-isolated technique of the Delphi-like method enabled a convergence that had not occurred in other forums, in which only one standpoint was present. By the end of this phase, discussion had already unanimously turned to identifying and developing projects that would decrease the environmental harms while safeguarding the economic gains of the e-waste industry.

The articulated development desires of the community representatives suggest that the central question for Palestinians is not simply how to stop Israeli e-waste importation, but to a large degree, how to selectively and carefully conduct a social and environmental upgrading of a recycling cluster to protect local lives, environment, and livelihoods. The West Line and similar e-waste hubs internationally all emerged in economically marginalized areas and are facing a legacy of severe contamination. If their economic foundation was undermined by eliminating, rather than building up, the sector, they would suffer a double devastation at precisely the moment when they most require the resources to move onward. This nuanced exploration of the development desires of an e-waste-dependent community adds to debates and discussions among the e-waste literature, and NGOs and policymakers who seek to improve the conditions of e-waste hubs. This elicitation of the West Line community's development desires and perception of imported e-waste as a development opportunity, if it is regulated and environmentally sound, introduces an often overlooked perspective to arguments focusing solely on the injustice of Global North–Global South e-waste trade.

With this overall community-defined development path clarified, the remaining phases of the Delphi method elicited assets at the community's disposal that could be leveraged to develop and implement specific projects to achieve these goals. In the first round of interviews, community representatives had no end to the list of projects they wanted to implement in their community, some of which included: conducting scientific studies analyzing the health and environmental damages of informal e-waste processing, developing and enforcing a basic license scheme for scrapyards, organizing an environmental police force to patrol and arrest those burning e-waste, researching business models to extract value from "non-valuable" scrap waste (e.g. certain plastics and foams), establishing a municipal waste dump, remediating hazardous burn sites (particularly those located near residential areas), conducting an awareness campaign to inform the community of the dangers of improper e-waste recycling, establishing an e-waste industry association to represent their interests to national governments, and even petitioning the Israeli authorities to remove hazardous waste from the e-waste sector, among others.

Apart from awareness-raising efforts as a means to achieve social change – which has been argued to be one of the least effective waste management strategies (Liboiron, 2014) – all selected projects required some form of outside support to succeed: financial, expert, or political. Some of the more ambitious projects requiring external support (e.g., efforts to formalize and regulate e-waste processing) leveraged the expertise and network of the "project consultants" already strategically embedded within the Delphi method. Several project consultants advocated for and spearheaded such projects, guided by and in continual dialogue with community representatives. The most notable resource we obtained was the financial and political support given by the Swedish Consulate in Jerusalem through the Swedish International Development Cooperation Agency (Sida), responsible for organizing the bulk of Sweden's official development assistance, which invited our submission of a proposal to fund community-defined projects that regulate and improve the West Line e-waste industry. Our submission reflected the development trajectory and selected projects articulated by the community, centering on the three most pressing interlocking components: (1) prevent open-air burning of e-waste; (2) remediate e-waste burn sites; and (3) sustain the e-waste industry by supporting a transition to a regulated and sustainable recycling sector. Given the complexity and ambition of the proposed project, Sida generously funded a series of preparatory grants to pilot and refine each element in the proposed package of interventions, and in doing so garner the support necessary for them from Israeli and Palestinian government branches.

Thus, the Delphi process identified a north-star development trajectory for the West Line villages consisting of three interlocked strategic pathways. The process of piloting, refining, and advocating for each component within the intervention package involved sustained engagement with the community representatives and project consultants to elaborate this framework into a detailed program for achieving the selected projects and mobilizing the resources and support to begin implementation. Using a template akin to the familiar logical framework approach to project planning, through extended communication rounds, we collectively identified the anticipated results of each project, resources needed, potential risks, and possible monitoring targets to measure progress. We detail the process and results of each component in Chapters 6–8.

Thus, the emerging results of this Delphi approach soon outweighed the theoretical qualms with which this chapter began. The approach allowed a heterogeneous and conflicted community to overcome a long-standing deadlock regarding the e-waste industry's health and environmental impacts, and converge on a shared development trajectory toward its regulated and clean operation, which could be specified into concrete plans that could be trialed through pilot projects and submitted as proposals for funding support. The fact that the overall thrust and package of proposed interventions endured over time, including through various transitions in the local authorities' leadership and structure, is a testament to the robustness of and broad support for the product of the process.

4.5 Concluding Remarks

This chapter introduced a nuanced perspective on the development desires of a community hosting an e-waste hub, with broader relevance for policy stances on Global North–Global South e-waste trade and e-waste hubs in the Global South. The West Line case underscores the divergence between the stance of national governments and NGOs, resonant with certain local civic groups, all distant from the economic benefits of the e-waste industry, and the widely endorsed development aspirations to regulate and formalize an e-waste industry that emerged from our noncoercive process of iterative inputs from diverse community standpoints. This emerging perspective contrasted with ones that had begun to circulate, which focused exclusively on the harm of the e-waste trade, framed participants as victimized marginalized populations dependent on foreign e-waste, and offered a policy recourse consisting of a sweeping ban on cross-border e-waste trade. The proclamation of a ban was bold, and, also, patently unachievable by the community or, indeed, other Palestinian actors, that would have either perpetuated the status quo of destructive recycling, or been achieved by outside forces in a way that crushed, rather than reformed, this important economic sector.

We believe that a different and more broadly representative and shared account of the e-waste problematic is possible in other contexts too, through similar kinds of analysis and community-driven participation procedures. At the same time, while we found the process we employed to be exceptionally productive, we concur with the caution of other scholars and practitioners working to reclaim community participation from increasingly routinized performance: this kind of effort is demanding and not without its own paradoxes and limitations.

In the next chapter, we think critically about the broader lessons of our community engagement in the West Line villages for e-waste management, elaborating a hub-driven approach to e-waste policies we think can and should be forwarded in other contexts. We suggest several converging rationales for e-waste policies that are centered around e-waste hubs, and present the carefully integrated hub-driven package of business, enforcement, and cleanup measures we forwarded in the West Line, as a model for hub-centered reform.

5 *Can Tails Wag the Dog?*

A Hub-Driven Approach to E-Waste Reform

In previous chapters, we discussed the emergence and circulation of simplified global narratives regarding e-waste hubs (Chapter 3) which have been embedded in the Ban Amendment to the Basel Convention and the Extended Producer Responsibility (EPR) model that have become templates for e-waste regulation and legislation. We saw how this kind of framing can become inscribed and solidified at the local level in national discourses and legislation.

In Chapter 4, we illustrated how this global "dumping" narrative can also come to frame local narratives and interventions through projects that ostensibly reach out to and partner with communities in the hubs in question. We also began to get a sense of how such hubs in the Global South – where much of the world's e-waste is processed – might push back against this kind of hegemonic conceptualization of e-waste originating in the Global North. We demonstrated an alternative participatory procedure in which researchers/participants assisted in the iterative elaboration of a broader and more consensual vision for resolving this e-waste hub's dilemmas. In this chapter, we situate this effort within a broader call for hub-centered e-waste policies.[1] We begin by showing that e-waste hubs figure centrally but, also, negligibly in the thinking of policymakers. We then elaborate on multiple ways in which the kind of hub-driven package of measures developed in the West Line offers a pragmatically and ethically superior approach to the e-waste dilemma.

After the sketching of the broader rationale of a hub-driven approach in this chapter, the three chapters that follow each describe a different key pillar of the integrated policy package formulated through our community engagement in the West Line. They consist of interlocked efforts to (1) build local community-supported

[1] This chapter is based on our publication submitted to *Environmental Politics* (Garb and Davis, 2023).

capacities and mechanisms to curb destructive practices; and, once contamination generation is reliably reduced, (2) remediate the environmental damages from decades of unregulated operation, through clearing contaminants, rehabilitating landscapes, and reducing exposure risks. These are predicated on (3) opening up a horizon of alternative livelihood options (emphasizing, but not limited to, facilitating environmentally sound modes of e-waste processing better interfaced with emerging EPR systems).

It is important to note that our articulation of the rationales of a hub-driven approach in this chapter, and of its key components in the chapters that follow, is done after the fact. We did not possess this template when we began, but became aware of the necessity of a hub-driven formulation of measures, of its three main axes, and of the importance of delivering these as an integrated package by trial and error, as we went. As we worked with the community to specify measures and pilot them, discussing and revising as we went, we began to discern the systemic logic of the parts and how they fitted together. Studying and dialoging with the efforts of others working in e-waste hubs worldwide suggested the robustness and broader applicability of this model. While evolved in a highly specific context, we believe that the package of measures we describe and the impulse behind them have structural similarities with other contexts with a prevalent informal sector, and have relevance for them.

5.1 The Centrality of E-Waste Hubs and Rationales for Hub-Driven Policy Packages

The e-waste problem has been the subject of thousands of academic publications with direct policy implications. However, while most e-waste likely ends up in informal processing hubs, these sites appear in the policy literature only as iconic markers of the problem (Davis et al., 2019; Radulovic, 2018), if at all. This is understandable: Peoples' experience, future vision, and description of the present are entwined and coproduced. Few of those in a position to write professionally about solutions for the e-waste problem are immersed on a regular basis in these end points or dependent on recycling for their living. Therefore, their attention and construction of problems and solutions will be more naturally focused by and through their own lived experiences and incentive structures. These privilege the contexts and formal systems that produce, collect, recycle, study, and track electronic and electrical goods; the distant informal world will appear not as offering alternative modalities and strengths but as suffering from a deficit of activity and formality – readily remedied through extension and expansion of formal systems.

It is not only within the e-waste stream of solid waste where the informal sector is not recognized as a suitable form of waste management infrastructure. For example, discussion of the circular economy has attracted growing interest within global policy and business in the Global North as an innovative approach to waste management that can increase the sustainability of production and consumption; yet the practice of circular resource flows has long been an unrecognized central function of informal recyclers around the world (Gutbertlet et al., 2017; Gutbertlet & Carenzo, 2020). For example, recyclers throughout the Global South reclaim a wide array of discarded materials ranging from paper and cardboard to plastics, metals, glass, and, of course, e-waste (Wilson et al., 2006). The amounts recovered are often unquantified, yet believed to be significant, capturing and processing most recyclable materials in the Global South (Dias, 2016; Kasinja & Tilley 2018; Kumar et al., 2018). In doing so, the work of informal recyclers provides environmental contributions and climate change mitigation by diverting recyclables from landfills and promoting transitions to sustainable economies.

Despite their contributions to solid waste management and circular economy transitions, most informal recyclers have no access to state-sponsored social protection, are exposed to occupational hazards and risks, and often face social stigma and economic exploitation. In most places, the contributions and insights of these recyclers are not formally recognized by local authorities when establishing new waste management systems (Gutbertlet et al., 2017; Gutbertlet & Carenzo, 2020), although nongovernmental organization (NGO) and academic partners have made strides to legitimize their contributions and demonstrate opportunities to tackle poverty reduction through government recognition and resources via inclusive waste management systems (Dias, 2016; Gutbertlet & Carenzo, 2020; Rutkowski & Rutkwoski, 2015; Velis et al., 2012). Circular economy policy initiatives in the Global North tend to frame waste management challenges as engineering and governance problems to redirect resource loops (Geissdoerfer et al., 2017). Yet, from a Global South and livelihood perspective, a critical issue of waste management includes the social inclusion and participation in public policies of the large population of marginalized recyclers (Velis, 2017). Gutbertlet and Carenzo (2020) argue that informal recyclers should be reframed as "major protagonists" in waste management, especially when they are organized in local and regional cooperatives.

These patterns resurface in the e-waste literature. For example, an April 2019 review of the 500+ papers in all disciplinary domains in Web of Science referencing electronic waste and policy (with variants thereof) in any database field shows only a handful that place the hub at the center of an initiative or policy. To give an example of the typical degree of invisibility possible within central policymaking arenas, a recent comprehensive review of stakeholder perspectives on EPR and the circular economy (Kunz et al., 2018) does not even include the informal sector and/or hubs in their list of actors: producers, Producer Responsibility Organizations (PROs), (formal) waste operators, national authorities, municipalities, and trade associations. Similarly, an earlier INSEAD report (Kunz et al., 2014) on the future of EPR and e-waste conducted interviews with twelve stakeholder categories, including "illegal informal sector" and "communities" (defined as "people affected by the negative externalities occurring at the end of life of electrical and electronic equipment," p. 18). However, neither of these latter categories were included among the interviewees of this comprehensive report. The report's description of the "illegal informal sector" indicates the mindset allowing this kind of omission:

> This group of stakeholders represents all private actors (companies or individuals) conducting collection and recycling operations illegally. This grey market developed out of the opportunity to generate revenue from WEEE. These actors are responsible for the leakages of WEEE outside official EPR flows, either by exporting WEEE illegally to developing countries, pilfering waste and reselling it to waste brokers, or recovering only valuable waste and landfilling the non-valuable part. This group of illegal actors goes against the fundamental objectives of EPR (p. 18).

It is only in recent years, and in the disciplines of anthropology, geography, sociology, and development studies, that we are beginning to get fine-grained portraits and novel perspectives emerging directly from the informal sectors and locations responsible for the majority of e-waste collection and processing in the Global South. For example, we have increasing recognition of the importance and value creation of scrap picking and e-waste recycling centers (Corwin, 2018;

Davis & Garb, 2015; Gidwani, 2015; Rathore, 2020; Reddy, 2013; Rochman et al., 2017), reflections on their emergence (Davis et al., 2019; Lora-Wainwright, 2016), dynamics over time (Davis & Garb, 2019; Reddy, 2015), and attempts to systematically quantify the flows and value added of industrial clusters whose informal nature requires novel methodological approaches (Amoyaw-Osei et al., 2011; Davis & Garb, 2018; Manhart et al., 2011; Ogungbuyi et al., 2012; Prakash et al., 2010).

The emerging consensus of this literature is that the informal sector and their focal processing hubs have evolved remarkable collection networks as well as applied knowledge of e-waste recycling economies. To survive, they have been keenly attentive to the changing flows of generated waste, the technical parameters and constraints of collection and processing, the realistic market forces, and the weak links and gaps in formal policies and practices. In moral terms, these places have a certain precedence. Over the many years before e-waste was "discovered" as a treasure or hazard, hubs in the Global South were an implicit element in the Global North's excess consumption, allowing developed countries the luxury of ignorance regarding the destiny of their end-of-life goods.

A growing recognition of the centrality of e-waste hubs, yet inability of policymaking processes to prioritize, mobilize, and authentically engage with them, has led scholars to reflect on national e-waste visions and policies along with the pathways and challenges to equitably reform e-waste management. In South Africa, for example, Lawhon (2013) describes two competing environmental discourses that underpin opposing claims of e-waste problems and solutions. The "ecological modernization" discourse adapts European lessons to build win–win systems that improve economic and environmental e-waste consequences whereas the "environmental justice" discourse advocates for the elimination of e-waste dumping in South Africa. Both framings offer a cognitive shortcut to understand complex e-waste realities but provide little utility when applied to South Africa and other middle-income countries. In such contexts, European recycling standards are an unrealistic short-term aspiration and e-waste dumping narratives, as discussed in the Israel–Palestine context in Chapter 3, are also overly simplistic. Despite these differences, productive engagements exist as e-waste governance proposals are driven by shared aspirational goals to increase recovery rates and cost efficiency while generating employment for marginalized populations (Lawhon, 2012b). Diverse stakeholders comprised of manufacturers, recyclers, civil society, and government generally agree that producers should be responsible for e-waste and an external subsidy is required for economically viable e-waste recycling. The details for e-waste governance systems, however, remain controversial, as three organizations – each composed of different stakeholder groups with their own claims to legitimacy (although the informal sector is unrepresented in each) – have developed pathways with differing priorities, funding mechanisms, and organizational structures to best manage e-waste. Lawhon (2012b) argues that the very presence of multiple visions and pathways for e-waste futures has produced contestations over power, trust, and legitimacy, which has ultimately divided resources, created confusion, and delayed e-waste management initiatives.

Smit and Musango (2015a, 2015b) also approach e-waste from an informal economy lens in South Africa, recognizing that the informal sector already performs green recycling work with considerable scope to expand through supportive integration into formal systems. Yet, recent policy questions regarding the proactive integration of informal recyclers remain unresolved as e-waste proposals insufficiently engage and mobilize key stakeholders (Lawhon, 2012a) and differ in whether the government or a nonprofit organization should support and facilitate such a transition (Grant, 2019).

Similar challenges have been described in Ghana. Oteng-Ababio et al. (2020) argue that Ghana's recent e-waste management policy – Act 917 – which, technically, embraces circular economy thinking and linking formal companies with informal operators (Grant & Oteng-Ababio, 2021), has comprised a "mishmash of techniques" combining development programs, eviction, and EPR-based regulation derived from a range of uncoordinated actors. Ghana's national legislation entails top-down management through strict official control to corral e-waste activities coupled with capital–technical initiatives supported by various international partners (Grant & Oteng-Ababio, 2021). Yet, beyond a few place-based initiatives, there is limited political will and creative imagination to engage and support informal recyclers. The inclusion and upgrading initiatives for informal operators are limited to them providing inputs to formal firms, one of the least lucrative nodes in the e-waste economy. Discriminatory attitudes toward informal workers, an absence of informal worker's voice in policymaking, insufficient funding mechanisms and capacity building to upgrade informal operators, and a general unawareness of informal e-waste activities and management have hindered their participation in e-waste management efforts (Chen et al., 2020; Oteng-Ababio et al., 2020). Oteng-Ababio et al. (2017) conclude that unless informal recyclers' livelihoods are recognized, valued, and accounted for in local economic development, e-waste formalization initiatives will continue to achieve limited success.

Grant and Oteng-Ababio (2021) contend that e-waste policy concerns in Ghana must focus on retaining informal operators' entrepreneurial attributes while facilitating conditions to encourage environmentally sound technology, improving incomes and work conditions, and being compliant with e-waste regulations. Considering Ghana's current socioeconomic development, it remains unlikely that the entire e-waste economy will operate formally nor will informal operators disappear. Thus, productive e-waste formalization mechanisms are argued to materialize through a continuum of "gray spaces" where informal workers ally themselves with governmental and other actors through cooperative agreements based on new formal infrastructure (Grant & Oteng-Ababio, 2021). To this end, Oteng-Ababio et al. (2020) call for increased policy coherence via a critical systems thinking (CST) paradigm for e-waste regulation by exploring polycentric tools where regulatory power arises from a series of constraints rather than being driven solely from the state.

The experiences in South Africa and Ghana outline how multiple e-waste visions and pragmatic approaches fragment resources and undermine policy coherence, and that e-waste hubs are paid lip service, but not considered as the "main protagonists" for e-waste reform. In some ways then, e-waste hubs are emblematic sites of boom and exploitation, abandonment, and reinvention of globalized capitalism – sites of *disposal* alongside the more familiar sites of exploitative *extraction* (mining, lumbering, drilling) and *production* from which wealth is built. For this reason, we liken them to company towns faced with either closure and abandonment or else some kind of more inclusive and constructive reinvention, in which they play a central role for ethical and strategic reasons. Because they embody both the livelihood and hazard faces of e-waste, e-waste hubs offer a position uniquely able to force an unfragmented and durable realignment of e-waste recycling networks. Conversely, their exclusion from policy decisions and formal recycling systems makes it more likely that the lure of improper practices will remain, merely mutating and diffusing the problem. For example, as typically happens, formal e-waste processes would continue to rely on and leak the lower value fractions to the informal sector, which would persist but in a further degraded standing.

Hubs are unifying of policy in an additional sense, in that they are also where the legacy of e-waste damage occurs and remains, and they are uniquely capable of addressing this. If health and environmental harm are the motivations for e-waste policies, then remediation of the continuing impacts of past practices is just as important as reducing future bad practices. Even if all flows to informal hubs ceased tomorrow through redirection to formal channels in a viable and hermetic manner, this would, at best, be a half measure. Most of the damage from improper e-waste disposal has been done and is lying on the ground continuing to disperse and poison, until cleaned. It is too tempting both cognitively and materially for Global North–driven policies to be forward-looking in their focus to capture current and future flows while implicitly jettisoning the trashed places upon which the electronic revolution was built. Because it is by and for the communities living at the junction of past and future damage, hub-driven approaches assure that both will be reflected in policies.

These, in brief, are the reasons for our claim that while hub-driven solutions are not, in themselves, sufficient, they are a necessary and catalytic part of global e-waste solutions. If e-waste problems are not solved for the hubs, they are not solved. Collectively, and in concert with variants of the familiar policies at the beginning of the e-waste pipeline, such packages can begin to redraw the e-waste economy in truly sustainable ways.

Despite its striking particularity, the Israel–Palestine context surfaces the exclusionary effects of EPR policies and the privatization of e-waste collection and recycling that occur globally – usually in places where colonial relations are less visible and immediate. Our research and activism within this context point to hub-driven policy packages as a necessary approach to e-waste reform. We find multiple converging rationales that help understand the importance of a hub-based approach to e-waste policy reform – each with associated disciplinary contexts and literature. In what follows, we draw on these to outline our moral and pragmatic rationales for placing e-waste hubs at the center of e-waste reform, and to outline what the contours of this might consist of.

5.1.1 Rethinking Governance in Waste Economies

As discussed in Chapter 1, around three-quarters of all postconsumer electronics travel through informal networks (Forti et al., 2020). This global pattern echoes in the Israel–Palestine context as the West Line e-waste hub has historically collected and processed between 42 and 54 percent of the estimated 120,000–138,000 tonnes of e-waste generated annually in Israel (Davis & Garb, 2018a). This is just one main site of e-waste processing and does not consider the activity of other mini hubs scattered throughout the West Bank or informal e-waste processing that occurs in Israel. Despite their numeric prominence, e-waste hubs that process most of the world's e-waste have had little say in shaping the legislation that is intended to govern e-waste markets and flows and reduce the risks of improper treatment.

Policies that have been formulated from the production side of things are less suited to the afterlife of the goods produced, which Gidwani and Reddy (2011) refer to as the "political other of capitalist value," a shadow domain of spaces, actors, and revaluation outside the pale of "production." Global recycling networks have a different logic, actors, and geography of flows than the systems conceived of from the traditional linear global production network or global value chain frameworks (Lepawsky & Mather, 2011), and thus require different governance constellations (Crang et al., 2013). Because canonical EPR e-waste policies were forged in a Global North–centered context – that is, in places without a robust and longstanding informal recycling sector – and in a manner that misses or misinterprets global recycling networks, they can be inadequate and harmful to the contexts of the Global South in which e-waste recycling largely occurs (Reddy, 2016). Because they center on the end nodes and, often, drivers of the afterlife of electronics, hub-driven policies offer a key leverage for a corrective.

5.1.2 Anti-Fragmentation: Hub-Based Packages Boost Policy Coherence

E-waste hubs, by their nature, draw together in a single place the spatial, temporal, and social elements enmeshed in e-waste issues, and are inhabited by communities that have dramatic stakes in both the livelihood and risk aspects of the e-waste equation. Their encounter with e-waste is not glancing, one among several diffuse abstract environmental issues, or through fleeting touches with products on their way to somewhere else. The consequences of e-waste are embedded in their landscapes, bodies, and household economies, so that their concern is uniquely and directly cross-generational. This kind of temporal sensibility and vulnerability lends itself to thinking about sustainability in a concrete way. In particular, it assures attention to the legacy aspect of the e-waste issue and promotes continuity of effort over time. This rare and valuable combination of the hub communities' urgency and sustained focus makes hub-driven interventions important to drive coherent systemic change. A solution translates into their children being woken by smoke or not; the topic is not one that can be picked up and put down, when convenient. In the West Line e-waste hub, for example, it would be hard to find a household whose extended family were not somehow engaged with or dependent on the e-waste trade for livelihood, or whose members had not been struck by cancer or a congenital disability seen to be associated with e-waste contamination.

Several bibliometric studies of e-waste literature (Andrade et al., 2019; Borthakur & Singh, 2020; Gao et al., 2019) consistently show its fragmentation among three disjoint communities of concern: e-waste management, e-waste toxicants, and e-waste value recovery. E-waste hubs, whose experience is direct rather than reflected in academic literature, grapple with all three aspects of e-waste in an intrinsically entwined manner. Further, the reuse and refurbishment dimension of waste, which often drops out of e-waste literature and policies (Gregson & Crang, 2015; Lepawsky, 2018, 2020), is more salient in e-waste hubs where both reuse and dismantling coexist. Thus, at a collective level, knowledge, incentive structures, and capacities for response will tend to be more robustly integrated with e-waste hubs and, conceivably, in policies and initiatives that emerge from them.

The kind of integration that would be boosted by a hub-driven approach has been forwarded in other policy spheres under different names: policy portfolios or policy mixes (Howlett & Rayner, 2007), policy coherence (Koff, 2017; Koff & Maganda, 2019), and policy packages (Givoni et al., 2013). These frameworks share the sense that measures intended to intervene in complex multi-actor and multi-scale realities fail when applied in an isolated manner because of unintended consequences, their inability to deal with institutional or partisan interests, and the cross-cutting and linked nature of the shifts required for change. Packaged co-deployment ("bundling") can harness latent interdependencies of individual measures that will boost the effectiveness of each, minimize possible side effects, and broaden coalitions and legitimacy. Our argument here is that e-waste policies have suffered from this kind of fragmented effect, in large measure because they ignore or poorly integrate the informal sector. Even policies with the explicit goal of integrating the informal sector have suffered from this kind of fragmented effect. A notable example is the "Best-of-2-worlds" (Bo2W) philosophy that was pioneered at a transnational scale by the Solving the E-waste Problem (StEP) initiative, which comprises a coalition of government agencies, NGOs, universities, and UN organizations. The Bo2W philosophy aims to technically and logistically integrate the best "pre-processing" capacities of developing countries with the best "end-processing" capacities of developed countries (Wang et al., 2012). However, implementations of the Bo2W initiative in China and India have yielded poor results due to a lack of deep cooperation with the informal sector (Rochat et al., 2008; Wang et al., 2012; Yang et al., 2008; Yu et al., 2010). Since these initiatives were made *for* rather than *with* e-waste hubs, the collection systems suffered from either limited participation or parallel competition that undermined their success (Davis & Garb, 2015). Packages of interventions emerging from hubs offer an elegant way of boosting the coherence of e-waste policies. As we describe below, the interventions evolved iteratively by the West Line hub community embodied a tightly interlocked and sequenced set of three kinds of measures, either of which alone or in a different order would have been ineffective or even destructive.

5.1.3 Hubs as Sites for Democratic Decentralization

There has been extensive examination of the interaction between destructive environmental processes and power and governance structures, the question of subsidiarity (which powers and decisions are best done at different scales and levels of governance), and debate over how to best lodge resources and authority at appropriate levels (for an overview, see Ribot, 2004). Many developing nations have endorsed the impulse for decentralization in various realms (fiscal, services, health, education), including in natural resource management (forestry, river basin management, biodiversity conservation). Arguments for decentralization include efficiency, democracy, inclusiveness, increasing participation, lessening political tensions, and downsizing to make central government more effective.[2] E-waste thinking and policies tend toward the national, with little exploration of the possibilities for decentralized measures. There are different phases of e-waste recycling, some lending themselves to decentralized operation and others less so: collection, aggregation, refurbishing, disassembly, metal extraction, disposal of non-valuable fractions, disposal of highly toxic fractions, and enforcement of clean operation. EPR tends to centralize all of these.

Collection (waste picking) is a key and dominantly informal phase (Coletto & Bisschop, 2017; Hayami et al., 2006), one in which attempts at organization are most advanced (Gutberlet, 2012; Kasinja & Tilley, 2018; Medina, 2007; Samson, 2009; Tirado-Soto et al., 2013) and institutionally embedded (e.g., the Global Alliance of Waste Pickers or efforts of Women in Informal Employment: Globalizing and Organizing, WIEGO), and in which the displacing effects of formal systems are well documented (Hartmann, 2018; Reddy, 2015). Similarly, in the dismantling phase, there are also advantages to the localized dismantling of many fractions and to localized enforcement against unsound practices (Davis & Garb, 2020). On the other hand, the waste fractions with negative economic value or high toxicity, or those that require cutting-edge technologies to be viable, are best handled in centralized facilities.

We suggest that e-waste policies should differentiate between measures in which a centralized approach has added value and be more flexible about dispersing governance and resources to decentralized sites when those have advantages. Currently, EPR policies ignore or implicitly – even covertly – rely on the informal sector and are thus handicapped in thinking about focused incentives and key points of intervention that will mobilize and build capacity of decentralized actors. The default orientation of hub-driven policies, unless there are strong reasons otherwise, is to lodge resources and decision-making power with governing bodies that are transparent, capable, and downwardly accountable to a broad base of local community members.

[2] We will not here debate the virtues or mechanisms of decentralization, nor the ways in which it can degenerate into damagingly incomplete decentralization, or merely deconcentration or privatization, which transfer burdens and costs without resources and authority (Ribot, 2003).

These downwardly accountable local sources of governance will have lots to say about the kind of central measures (e.g., inspections at scrapyards or ports) that would serve their abilities and efforts to upgrade.

While the Palestinian and informal e-waste sector context differs from the rural development literature in which much of the democratic decentralization debate has emerged, some key dynamics and insights support hub-driven reform. Specifically, processes and resources of hub-driven policies are more likely to empower and strengthen local downwardly accountable actors. This is especially important in e-waste hubs, which tend to emerge in regions with ambiguous administration and governance vacuums (Davis et al., 2019). In the Palestinian context, several factors have conspired to weaken local governments, whether through the upward allocation of authority and resources, the inability of governing areas fragmented and constrained by the Israeli occupation (especially in Area C), corruption and nepotism, and the saturation of development and services sector by donor and NGO-driven initiatives, which tend to sideline the elected and accountable structures.

5.1.4 *Informal Innovation and Ingenuity*

Seen from the perspective of formal systems, informal e-waste businesses seem marginal. When defined by comparison with large capital and technology-intensive facilities and operations coupled with global expertise and perspectives, they seem second-best. Yet looked at not through the lens of absence but through the distinct characteristics of the informal way of doing things, there are many spheres in which hubs are positioned to facilitate innovation, agility, specialization, and economic viability, especially when it comes to operating in shifting local contexts (Kraemer-Mbula & Wunsch-Vincent, 2016). The challenge for e-waste policies is in sorting out the spheres in which formal and informal skillsets and orientations have the most to offer.

Obviously, there are kinds of recycling that should only take place in sites in which the economies of scale, resources, and ability for research and development (metallurgy, robotic dismantling, etc.) available to the formal sector are present. On the other hand, informal systems seem to perform at great advantage in the collection phase of waste treatment and boost the reuse and refurbishment dimension of waste.

Reuse and dismantling coexist in e-waste hubs, and the low price of labor allows a more nuanced approach to waste streams. A large factory that treats waste on a conveyer-belt basis might simply compact printers into low-value blocks of mixed plastic and metal rather than test and dismantle them in a way that can extract greater economic and environmental value. Because the informal sector operates without subsidies or external financing and is remote from opportunities to engage in rent-seeking lobbying of policies and regulation, it must operate in a way that is keenly and realistically attuned to market dynamics and opportunities, and especially to locally specific ones. Having skin in the game and autonomy encourages a stance that is simultaneously more risk-taking and more conservative – both in a useful sense. Giving hubs an increasing role in shaping policies might allow more opportunities to cultivate and express these advantages.

For example, in the West Line e-waste hub, we find businesses dedicated to and developing expertise in niche waste streams that would not be handled as resourcefully in large recycling facilities: a single person specializing in refurbishing gym treadmills or electric generators, for example. Another business owner, an electrical engineer by training, has reverse engineered and reprogrammed the embedded controller of a used cable grinder he purchased from Turkey in a way that allows considerably less loss in handling heterogeneous, variable loads of cables from scores of dismantlers. In fact, the large Chinese manufacturer of these machines has approached him to barter this innovation by trading it for a massive reduction in the price of new machinery. Waste streams that would be simply compressed into mixed plastic/metal cubes for shipment by a large formal facility can be hand-dismantled in an informal one, with the separated streams having far greater value than the aggregate. New waste streams or single lots that could not be handled by fixed processing chains can be tackled.

While nimble and able to experiment, hubs are also likely to be hard-headed and conservative in their evaluation of feasibility. They are not buffered by the moral hazard of processing e-waste nor capital reserves of other actors, which can make decisions about how much risk to take without bearing the cost if something goes wrong. We have watched local scrap businesses nod politely and then disregard recycling schemes proposed by outside experts and NGOs. Donors or remote ministry officials with no skin in the game can afford to be convinced by colorful graphics and wishful business models, but not those who will be there for the long haul. Their interest in a development project, for example, may reside in only the 10 percent of the project they know will remain long after the development brokers, with their overseas consultants and rented offices, have disappeared: a high-amperage electrical connection to their village, for example.

5.1.5 Cultivating Hubs as Industrial Clusters

What we have referred to as hubs are, in fact, informal economic/industrial clusters. Conventional descriptions of industry clusters derive from Alfred Marshall (1920) and emphasize three factors: the growth of intermediate and subsidiary industries that provide specialized inputs, a pool of skilled labor tailored for local industry, and the establishment of dedicated infrastructure and other collective resources. These three factors generate agglomeration economies, which create cost advantages for individual firms because of their location within a cluster of industrial growth, and synergies for the hub that benefits from the density and variety of their copresence. Because of these economic advantages, formal industrial clusters regularly benefit from government investments and resources to cultivate growth. Yet, little attention is given to cultivating and upgrading informal clusters, which hold similar attributes and can be more deserving of support than their formal counterparts.

In the West Line villages, the "core intermediate firms" are e-waste dismantling scrapyards, and a range of subsidiary businesses have emerged to service their needs or enhance the value extraction from e-waste. For example, refurbishers specializing in certain products (e.g., refrigerators, air conditioners, washing machines, etc.) will purchase functioning or repairable components of these items from dismantling facilities for a higher price than the raw metals alone, enabling dismantlers to increase profit margins with little added effort. Niche recycling facilities have also emerged to profit from low-value materials that dismantlers would otherwise discard. For example, several plastic recycling facilities have invested in grinding machinery and collect, sometimes free of charge, discarded plastic materials from scrapyards which are then ground into pellets and sold to local manufacturers. Since the selling price of plastics is quite low, sometimes fetching just ₪200 ($58)/tonne, plastic recyclers could not survive if they had to bear the cost of collection from Israel or even domestically, but are economically sustainable when inputs are sourced locally. Moreover, e-waste dismantlers avoid paying to dispose of unwanted plastic remainders.

Other businesses have opened to meet the needs of dismantling facilities. For example, a surge in local metal traders emerged to service smaller household scrapyards unable to transport metals directly to exporters in East Jerusalem, offering competitive pricing and a convenient outlet to sell extracted metals. Informal landfills provide dismantling facilities an economically appealing option to dispose of nonvaluable discards. Even several restaurants have opened in areas dominated by scrapyards to cater to the needs of traditional family-based Palestinian employers that must provide lunch for their workers. Other convenience stores have opened near dismantling facilities selling grinding plates and oxygen tanks for blowtorches required for certain dismantling processes. These supplemental businesses encourage clustered growth patterns by increasing the profit margins of scrapyards or making certain aspects of the dismantling operation more logistically convenient.

Scrapyards have also relied on a deep pool of teenagers and young adults in the area with expertise in dismantling e-waste. While this may sound like the glorification of a job that essentially smashes electronics to extract metal components, knowing exactly where to strike electronic items with the necessary force and knowing the right tools, requires practice and experience. Our own embarrassing attempts to dismantle e-waste items attest to the difference in skill level between a novice dismantler, which took members of our team the better half of an afternoon to dismantle one air conditioner, and a journeyman dismantler who can extract and group together each valuable component of an air conditioner in twenty minutes. Indeed, one of the primary growth constraints of dismantling facilities is finding workers who can dismantle e-waste and scrap metal quickly enough to earn the owner a profit.

Palestinian decision-makers and donors have devoted large efforts to build and support other economic sectors of far smaller scale. Therefore, it would make sense to consider rehabilitating the e-waste sector by improving its capacities for clean and increasingly formal functioning, rather than embarking upon a strategy of criminalization and attack. In fact, an initial adversarial stance from Israeli and

Palestinian governments did not eliminate the industry and achieve environmental and health gains, as hoped, but degraded environmental capacities and an appetite for improvement (Davis & Garb, 2020).

While many celebrated agglomeration benefits and entrepreneurial vibrancy of industrial clusters may well apply to e-waste hubs, some studies have shown ways in which informality can curtail the benefits of agglomeration economies in informal clusters. The social nature of skills acquisition and minimal financial investment tends to lower entry barriers in informal clusters (Ali et al., 2014), which can lead to congestion and high rates of competition that limit agglomeration benefits and can even motivate firms to relocate away from industry clusters (McCormick & Oyelaran-Oyeyinka, 2007; Sonobe et al., 2011). However, exiting informal clusters can be difficult as a lack of alternative development coupled with limited skills can lock in young laborers (Basole, 2016). The emergence and growth of informal clusters often occur either without the government's knowledge or despite government opposition. This limits participants' ability to draw on state support, and, in fact, government hostility can limit the benefits of agglomeration when informal clusters are perceived to be environmentally, legally, or economically problematic (Meagher, 2007; Mottaleb & Sonobe, 2013). Our own work has shown that limited access to capital for expansion or improvement (e.g., business loans) is a barrier to long-term thinking and to expansion and improvement. Because of these barriers, informal clusters might form and accelerate rapidly but reach a ceiling in their ability to evolve into higher-order industrial districts with robust and internal agglomeration economies (Harris, 2014).

However, despite the considerable academic and policy attention given to incubating and developing formal industrial clusters, with considerable state resources devoted to this end, there is almost no attention given to the equivalent cultivation and upgrading of informal clusters, which are especially needful of such support. In fact, in the case of the e-waste industry, the opposite is the case, as e-waste policies in their current form emphasize relocating e-waste flows away from existing hubs to remote formal businesses and locations, thus undermining the agglomeration of entrepreneurship and skill that might be celebrated and strengthened in a formal cluster. Often, discussion of industrial cluster upgrading emphasizes the sheerly economic dimensions of improvement at the risk of ignoring or even compromising the long-term social and environmental conditions of the cluster communities (de Oliveira, 2008). In the case of e-waste hubs, the social and environmental dimensions – such as improving environmental functioning, occupational health and safety, labor practices, and tax contribution – should be central to upgrading. An alternative to EPR policies that drain the economic lifeblood of hubs would be policies that infuse small local businesses with technical best practices and expertise, standards and enforcement, and increased linkages to formal actors, all done in a supportive rather than punitive manner, which recognizes the long-term trajectory of change.

5.1.6 Hubs as Restructuring Company Towns

We have just argued for the utility of framing e-waste hubs as industrial clusters, which merit the kind of cultivation and upgrading directed at their formal counterparts. At the same time, a similar framing and granting of agency is important in facing the harsher aspects of change or even downsizing that may be needed to improve these hubs in the long term. Not all businesses and e-waste fractions will be able to pass the threshold of clean operation that state-of-the-art facilities, typically in the Global North, can reliably offer. Environmentally adequate extraction of metals from motherboards, or the recycling of batteries and cathode ray tube (CRT) monitors, for example, is unlikely to be viable in facilities that are early on the path to formal and clean operation. Thus, there will be a contraction and restructuring aspect of hub upgrading, alongside the upgrading one. This, too, can be hub-driven.

We can draw here from the lessons of plant closures and structural change in company towns. Though e-waste hubs are composed of small and typically family-run businesses, rather than a single large company, there are parallels with the situation of a company town with a mining or chemical production facility that must close or relocate due to changing circumstances or unwillingness to tolerate the harms of operation. As in the classic company town situation, residents of the e-waste hub benefit from and are engaged in the production activities that harm them, and the town's economy is dependent on a single kind of production that is forced to shut down or change as its damages or unsustainability become apparent. Therefore, we can draw from the literature examining constructive pathways for plant closures and community restructuring strategies, which underscore the need for and benefits of a proactive partnership between the departing firm and the community.

Wigblad (1995), for example, describes a "community bridging" or "community restructuring" model, arguing that for a turnaround to succeed the closing strategy must be integrated with the community's growth effort. He urges that the departing business firm take this on. Timing is crucial. Decline must be in parallel to growth in new employment, so shutdown must be slowed or delayed until alternative options are in place and the new income starts to ramp up. Stakeholder cooperation is also essential, and advance warning must be given. In the classic case, the stakeholders would include senior management in the company, national government (e.g., to grant a tax break to encourage new facilities), local government, and the trade unions. For example, a downsizing company can help establish a community development corporation or offer a mixed package of measures (early retirement, encouragement of other companies, transitioning to new jobs, etc.). In this way, a proactive, forward-looking approach yields a more robust and sustainable outcome in the long term than a preservation strategy of marginal adjustments. The transition will require patience and, at times, short-term sacrifices for long-term gains. While the details of the community restructuring cases are clearly different in informal hubs than the ones discussed in these studies, the important thing is the heuristic of framing the situation in terms that draw on analogous knowledge and insights – in particular, the emphasis on the need for proactive partnership in the face of necessary industrial change. This approach will be impossible – indeed, invisible – if the policies driving these structural changes do not include hubs as major players.

5.2 The West Line Package

The above insights and claims about the need for and possibility of hub-driven realignments arise from nascent proposals to reform analogous e-waste hubs coupled with our experience in iterative community-based development, coalition building, and advocacy discussed in Chapter 4, and the subsequent testing of a package of measures described in the following three chapters. Here, we briefly summarize the three interlocked pillars of the West Line hub-driven policy package to demonstrate the extent to which this kind of approach might forward the hub-driven approach to e-waste policies whose moral and pragmatic rationale was sketched above. While this policy package has been extensively vetted and successfully piloted at the hub level – and advanced to a wide range of stakeholders at the level of principle – its full implementation has not yet been achieved due to the geopolitical complexities and power contestations in the Israel–Palestine context, which we discuss in the following three chapters. Thus, many aspects require development and clarification, especially the outward-directed aspects of how hub-level efforts should and can be linked to cooperation and reform measures beyond the hub. The following is a brief summary of these three central measures.

5.2.1 *Curb Environmentally Destructive Practices*

The most environmentally destructive practices in e-waste hubs, which have given urgency to formal e-waste management, must be directly rectified through community-supported mechanisms. Otherwise, they will continue unabated. In the West Line villages, the most salient and consequential environmental and public health harm is e-waste open burning. About 80 percent of e-waste burning is to extract valuable copper from plastic insulated cables, and the remainder is an inexpensive way to dispose of non-valuable waste (e.g., plastic, foam, glass) (Davis & Garb, 2020). This burning is the most proximal problem that propels local communities and adjacent Israeli communities to lobby for change in the sector. These environmental affronts can and must be reduced immediately, alongside longer-term efforts to transition an entire industry (as described below). This requires a coordinated multipronged set of interventions to address the diverse kinds, motivations, procedures, and e-waste burning locations. Our engagement with recyclers in the West Line villages helped us identify and pilot a series of initiatives that offered a carrot (an initially subsidized cable-grinding service as an economically attractive alternative to e-waste open burning) and then a stick (community-based enforcement against those persisting to burn) (Davis & Garb, 2020). During a three-month trial period, nearly three tonnes of cables were ground daily under this arrangement, drastically reducing the amount of pollution emitted in the West Line villages, with locals reporting an 80 percent reduction in burn events. In parallel, we began efforts to reduce the remaining 20 percent of e-waste burning done to dispose of non-valuable e-waste through machinery to valorize these waste streams, more reliable solid waste collection services, and affordable disposal options. Similar combinations of incentives and deterrents can help reduce other harmful practices besides burning.

5.2.2 Heal the Toxic Legacy

After years of "out-of-sight, out-of-mind" operation, most of the damage of an e-waste hub has already been set in motion through the release and persistence in the landscape of massive amounts of toxic substances. Stopping destructive practices is a critical first step, but this alone is not enough; it must be accompanied by massive cleanup and remediation of past contamination sites as well as heightened ongoing attention to the health of a generation of people exposed to dangerous contamination. For example, in the West Line villages, we identified over 250 e-waste burn sites present up to 2011 and showed the high concentrations of dangerous heavy metals and organic compounds in these sites and their gradual dispersal from them (Davis & Garb, 2018). The number of such sites that has more than doubled in the decade since. A thorough cleanup of surface ash and contaminated soils at the emission sources would go a long way toward lowering human and ecosystem risks. Our efforts to do so in the West Line villages included a systematic mapping of the burn sites; discussion of transparent selection criteria for which burn sites to prioritize, shared with and agreed upon by relevant local stakeholders; costing; and assessment of logistic procedures and options, with a special emphasis on the employment of local expertise and labor. We also attempted to boost awareness of and resources for the unfolding medical consequences of past exposures.

This cleanup also helps prevent the recurrence of burning, which tends to take place on land in which ownership is absent or lax, or, in some cases, where the owner explicitly obtains revenue from allowing burning. Once land is degraded by burning, a vicious cycle of dereliction can occur. A process of clarifying and valorizing land ownership, boosting the status of communal land, and conditioning cleanup on the willingness and ability of owners to prevent reoccurrence will increase the barriers to local burning.

5.2.3 Industry Upgrading and Gradual Formalization

Upgrading and gradually formalizing e-waste hubs requires efforts at multiple levels. Similar to the industry organizations representing businesses in other sectors, an "industry-voice" of Palestinian e-waste recyclers must emerge and forge a unified lobby and articulation of their collective interests. An assessment of e-waste items processed coupled with an economic and technical triage should be conducted as the basis for change efforts. This would identify the waste streams that are: (1) easy to process using off-the-shelf technologies and clear economic viability – cables, for example; (2) impossible (low or negative value and demanding separation technologies – e.g., CRTs or batteries); and (3) those in the middle, where bridging investments could possibly make profitable and clean recycling possible (refrigerators and air conditioners are probably in this category). Then, for the most viable segments first, the relevant technical best practices, flow volumes and pricing, and local buyers, and regulatory standards on both the Israeli and Palestinian sides, as well as the tricky question of cross-border transport under the Basel Convention, must be considered and debated with widening circles of stakeholders. It is critical that this is done in the spirit of finding and building the robustness of transition pathways, which might initially fall short of the formal letter of the law. Such pathways must not just offer an end goal, but assure businesses embarking on this route a reliable horizon of legitimacy that they need to invest and transition.

A key aspect of transition is creating a trading zone across the seamline between formal EPR systems and informal networks. As suggested by Inverardi-Ferri (2017), upgrading or transitioning recycling facilities can be subsidized through EPR funding mechanisms already in place (which are typically extracted through "end-of-life treatment" fees applied to the purchase of new electronics). Currently, these are skewed to direct the funds mobilized through EPR to large-scale formal recycling companies, which effectively compete with the informal sector. Instead, opportunities could be sought to use these funds for reforms that upgrade informal systems, rather than undermine them. This is not simply for the informal sector's benefit: PROs within an existing EPR system can obtain a massive boost in their collection capacities if informal flows can be registered within their system. To do so, they must work with informal traders' needs and provide assurance that such partnership is not simply a step in their slow usurpation (e.g., by gradually identifying and assimilating their suppliers). On the other hand, if informal recyclers work toward reliable certification of their environmental and worker health standards, then the formal system can send materials to the recyclers with confidence. Informal businesses will be willing to go that route to the extent they increasingly gain confidence that certification will link them to additional and stable inputs and help them avoid penalties (=costs) of informal (=illegal) operations.

The government can facilitate this partnership by offering flexibility in standards during a transition period and offering loans and a period of initial tax breaks to ease upgrading. Tax breaks (on income and value added tax) are a compelling and low-cost incentive for governments since the taxes waived are not ones that have ever been paid to the government by informal businesses. Governments and the EPR system can also buffer and bridge between promising and problematic waste streams in ways that individual businesses cannot, for example by cross-subsidizing between high- and negative-value materials, or by assisting businesses and households whose livelihoods may be negatively affected by the sector reform (e.g., cable burners or small recyclers that process highly toxic e-waste items).

5.3 The Limits of and Necessary Supplements to Hub-Driven Measures

We have presented the rationale for and contours of a hub-driven approach to reforming national e-waste policies through an integrated package of measures that emerge and radiate out from an e-waste hub to the broader contexts of e-waste management, including formal EPR policies. Our confidence in this approach derives from the converging theoretical grounds for its validity and from our practical experience in developing its details. These include the careful community-based evolution and piloting of the measures and the extent of buy-in from an unexpectedly broad range of stakeholders, including a cautious development donor willing to provide multi-million-dollar funding for full implementation. At the same time, we must moderate, qualify, and caution against overreach. Implementing such a package is a slow, rocky path, and we should not forget the liabilities of the informal sector as we celebrate its virtues.

In our case, despite acceptance by a range of Palestinian community, business, and municipal actors, and support from the Palestinian Prime Minister's Office, and relevant Israeli stakeholders, the project is currently frozen because of resistance at the senior levels of the Palestinian Environmental Quality Authority (EQA) needed for donor approval. This stasis revolves around political and jurisdictional tensions, as the head of the EQA was unwilling to approve any bilateral coordination with Israel prior to the resolution of the broader political conflict, and was loathe to devolve resources, decision-making, and enforcement to the municipal level at which these would be most effective and accountable. Hence, the kind of clarifications and testing of broader implementation, beyond the successful piloting stage, still lie ahead.

Also, while the moral and functional logic of a hub-driven footing for e-waste policies is strong, the informal sector has its own liabilities and biases, and hub-driven measures alone are not enough to drive the needed levels of systemic change. The unique leverage and potential of e-waste hubs for driving deep and coherent systemic change must not ignore the hazards and partiality of that standpoint. Flexibility and informality have a shadow side. When labor is cheap, so are lives, and some of the "competitiveness" and agility of hubs lie in this. Among the workers we interviewed, very few did not bear some scar of their occupation. And in the absence of outside professional enforcement of standards, locals are left to intuitive assessments of risk, which tend to normalize everyday exposures until it is too late. We also saw how readily externalization of hazard is accomplished in these settings. An informal landscape is one into which a troublesome load of discarded plastic can "disappear" into a burn site without a second thought. The same ingenuity that can quickly extract copper wire from electric motor armatures can also feed plastics with high concentrations of lead and flame retardants into recycling waste streams used to manufacture sandals or the springy matting for children's playgrounds (Mao et al., 2020).

Decentralization has compelling advantages to the extent it lodges decision-making power at the level of those who are downwardly accountable to those closest to damage. But does this make an e-waste cluster better able to self-regulate than the formal mining industry or chemical manufacturer? In some ways, yes. Our claim is not that the local level is immune to being swayed by proximal interests, but that in this they are not necessarily worse and may at times be better guides than the apparently global and disinterested motivations of remote decision-makers. Indeed, local communities can be fragmented with multiple cleavages along the lines of gender, class, clan, and positioning with respect to livelihood base (Davis & Garb, 2017). But so are remote decision-makers.

In this case, each of the remote decision-makers or advocates who would shape e-waste policies has their own agendas. The private sector would like to redirect the "leakage" of precious materials away from the informal sector into their own facilities. The Israeli Ministers of Environment of the last several administrations have been political appointments from militantly right-wing parties and have leaned toward using the potential cross-border consequences of Palestinian contamination as an argument for extending Israeli environmental law in the Occupied West Bank as a "soft" leading front of broader annexation initiatives. For its part, the Palestinian EQA has sometimes seemed more motivated to use the harm and destruction caused by the issue to showcase Israeli oppression than it has been to take a combative stance toward local populations in remote and already restless parts of its jurisdiction. The Hebron Governorate and the Civil Administration who must manage the daily needs of the occupied population are already chronically overloaded and would prefer to turn a blind eye to polluting activity if it provides livelihoods to already restless populations. NGOs find e-waste a sexy issue to harness for fundraising and relevance, and even academics, such as ourselves, can utilize the issue for their own external purposes, including serving as the basis for learned disquisitions such as this. In all this tangle of interests, it seems best to err on the side of those who are most directly affected: citizens whose children will have food on the table and asthma, or not, and mayors who know that sooner or later they will not be reelected if they do not solve the real needs of those they represent.

5.4 Concluding Remarks

We have presented a series of rationales for a hub-based approach to global e-waste sector reform, based upon an integrated package of measures arising from and driven by the hubs where most e-waste is processed. It should be emphasized that we are suggesting something more than simple additive inclusion of the voice of the hub community as another voice among stakeholders. We argue that an explicitly hub-driven approach, one in which the derivation and implementation of measures are built out from hubs, allows – indeed, demands – important new dimensions lacking in current e-waste policies. Such packages can begin to redraw the terms of the Bo2W philosophy, which has gained traction as a promising means to develop alternative recycling paths for domestic e-waste in developing countries. The Bo2W philosophy is not so much flawed in principle, but rather limited by the asymmetries inherent in its conception and execution. One cannot speak of "the best of two worlds" when each possesses different bargaining power in this fusion, resulting in a policy that acknowledges two worlds through the lens of one, and thus poses a transition that consists of the assimilation of one to the other (Lepawsky et al., 2017; Reddy, 2016). Until now, this has been the case, with more energy directed to shifting waste streams *away* from e-waste hubs to the large facilities where the tools and expertise reside than to working with hubs to adapt and bring best practices, knowledge, and tools *to* them. As we show, a hub-driven vantage point is likely to improve the temporal sweep, robustness, equity, holism, efficiency, and knowledge-base of e-waste policies in ways likely to make them far more sustainable and just.

Despite the multiple rationales we presented for a hub-driven approach to e-waste reform, there are obvious limitations and roadblocks. Within the Israeli-Palestinian case, we have argued that various stakeholders external to the West Line e-waste hub would benefit from such an approach: PROs and the Ministry of Environment in Israel can readily achieve e-waste collection quotas, the Palestinian Authority can economically benefit from a formal e-waste industry, and a reduction in harmful e-waste processing would generate wider environmental improvements. Yet, these ostensibly obvious mutual benefits have been overshadowed by contestations over resource allocation, decision-making power, and geopolitical tensions. As was evident in the South Africa and Ghana cases, without unified coordination among multiple stakeholders, persistent advocacy, and supportive external funding – which was atypical in the West Line case – e-waste hubs will struggle to achieve the political leverage and power necessary to shift established "best practices" for e-waste management to hub-driven alternatives.

This brings us to the second part of the book. In what follows, we detail our efforts with the West Line community to implement a hub-driven package of measures to reform their e-waste hub. By embracing the power and relevance of pragmatism – the belief that we can best know through doing (Brydon-Miller et al., 2003) – our efforts to change the Israel–Palestine e-waste system expose the deeply embedded structural forces that facilitated the emergence of this industry in the first place. In addition, we surface the power imbalances and political impasses between Israeli and Palestinian governments and within Palestine between municipal and federal government authorities that limited the successful implementation of these initiatives.

PART II

Pathways and Predicaments

6 E-Waste Burning

Causes, Consequences, and Corrections

Muhammed is a trained electrician, but at the age of twenty-five, like many Palestinians his age, he finds himself jobless. Regular employment is hard to find, and occasional work is sporadic and low paid. He is not one of the luckier few who obtain occasional permission to work in Israel, where jobs are plentiful and salaries can be four times as high, but he sometimes crosses over illegally to find work as an electrician. In the West Bank, the most profitable work for him by far is burning cables to extract and sell copper.

Muhammed is emblematic of the niche of burning and burners. We met him in 2012, early on in our fieldwork in the region. He arrived during an interview with a scrapyard owner to purchase 500 kilos of cables extracted from dismantling various discarded electronic equipment. We watched his heated negotiation with the scrapyard owner, revolving around their respective assessments of the cables' copper content and the current and projected price of copper on the international commodities market. In scrapyards such as this, which extract heterogeneous cables from different dismantled items, each batch is unique, so there is always room to debate a fair sale price. For example, at the time, "thin" cables found in most electronic devices (e.g., extension cords, computer chargers, etc.) typically sold for ₪7,000 per tonne, whereas "thick" cables from the electrical grid and elsewhere sold for a higher price of ₪10,000 per tonne since they contain a higher quantity of copper. The weight of plugs and other low-value attachments on the ends of cables also figured in negotiation, as these must be laboriously stripped to speed up the burning process and avoid contaminating the copper output. In the end, both parties agreed to ₪3,000 ($923) for the lot.

Muhammed was amused by our interest in this work and playful participation in the negotiation and so invited us to join him in preparing and burning the cables. We gladly accepted and followed him to his nearby home in Beit Awwa where we unloaded his recently purchased cables. These were spread onto a large plastic tarp and prepared for burning by using handheld cable cutters to remove adapters, plugs, battery packs, and other chunks of hard plastic that slow down burning and contain little metal. We then loaded the cables onto the truck and played cards and smoked shisha in the local café until it was dark enough to drive to the burn site undetected.

As evening arrived, we were joined by Muhammed's younger brother, and all packed into a three-tonne open-back truck and headed to a well-used remote burn site on the outskirts of Deir Samet, strategically nestled in a deep valley among rolling hills that minimize the visibility of the smoky process. To reduce exposure to the toxic smoke, we purchased masks from a local pharmacy and insisted on their use, a novelty that Muhammed and his brother found amusing, and donned. After a fifteen-minute drive winding through rural homes on a road wide enough for the passage of just one car, we arrived atop a hill, at the last intersection before the burn site, about a kilometer away. Soon, we were approached by the young son of the burn site's landowner, who functioned like a parking lot attendant, collecting the ₪200 fee and directing us to the designated burn site. The owner had "rented" his land in this way for years, affording burners a stable and hassle-free burn.

We descended the rugged dirt road toward the valley floor, the last few hundreds of meters of which were covered in a thick black ash layer, emitting mind-numbing levels of the noxious chemical smell that over the years we came to refer to, jokingly, as "the West Line smell." The half tonne of cables was unloaded from the truck, separated into five 100-kilo piles for faster burning, doused with gasoline, and lit with pieces of cardboard. In a few minutes, the cables were ablaze, sending billowing pillars of black smoke into the sky, a signal for a dozen or so nearby children and teenagers who descended the valley slope a few minutes later to be ready to collect copper fragments left behind by combing their fingers through the ash.

With the cables now fiercely ablaze and with work to do, our hosts stripped off the annoying protective masks and began stirring the fiery tangle with long iron rebar rods to ensure rapid and complete combustion of the cable insulation (see Figure 6.1). After forty-five minutes, with the plastic gone, the fire died down and Muhammed dumped buckets of water on the remaining copper to cool it off so we could load it onto the back of the truck. The 200 kilos of copper extracted

Figure 6.1 Muhammed's younger brother spreads around a pile of cables with a long iron rebar to speed up the combustion of the plastic insulation.

from a half tonne of cables (a typical 2:5 ratio) was later sold by Muhammed for ₪6,000. After the deduction of various production expenses, this left ₪2,000 ($615) for him and his brother – the equivalent of a good weekly salary in Palestine, for only a day or two of work. As we ascended from the burn site, three separate pillars of smoke were visible from adjacent valleys. The night was young, and the work of cable burning had just begun.

Muhammed's open-air e-waste burning is a variant of similar practices that take place all over the world, and the topic of this chapter. We begin by discussing the difficulties of policing this harmful but alluring practice, detailing the largely ineffective earlier attempts and the evolution of our own interventions in this regard. These consisted of both positive inducements, such as a non-burning means of metal extraction and waste disposal and the redirection of economic incentives to make these alternatives more feasible, and the deterrent measures of policing and enforcement, which might follow once non-burning cable recycling paths are available. In particular, we describe our pilot program, which embodied these approaches and achieved a drastic reduction in burning levels during its operation. We end with a snapshot of the current situation, in which tensions between local- and national-level forces have stymied the operation of what was shown to be a feasible and effective approach.

6.1 The Barriers to Burn Prevention

The plume of smoke we saw that day, towering up from a burn of 500 kilos of cables, was alarming, yet our subsequent research showed that it represented only 5 percent of the ten tonnes or more of cables burned daily in the West Line villages during the peak years of the industry's operation.[1] In addition to this "extraction burning" of cables is the "disposal burning" of non-valuable materials discarded from e-waste dismantling processes, such as the hard plastics and foams that have no economic market locally. Beyond e-waste, regular household waste, containing plastic and other potentially noxious materials, is also often burned in the places or times when solid waste collection systems falter. Our systematic mapping of the appearance and activity of burn sites since the early 2000s has identified hundreds of burn sites in the West Line villages, scores of which have been used in the sustained intensive manner as the one we observed with Muhammed.[2]

Cables and other electronic equipment contain various dangerous heavy metals (lead is a major risk, as well as arsenic and others), while the plastic coating and materials contain additives, such as flame retardants. When this mixture is burned in an uncontrolled manner in the open air, at a relatively low temperature, the incomplete combustion generates a variable and usually toxic mix of materials, which have demonstrated acute and chronic health effects including carcinogenic and mutagenic outcomes (Brigden et al., 2005; Gullett et al., 2007; IARC, 2011). These toxins are transported in fine particulate

[1] This estimate was derived from interviews with the two East Jerusalem metal traders (described in Chapter 2), who are the principal buyers of all burnt copper from the West Line villages. Combined, these two traders purchase over five tonnes of copper from the West Line villages daily, which would require the burning of at least ten tonnes of cables to produce. Anecdotal reports from others familiar with the industry within the villages confirm these numbers.

[2] An earlier discussion of the barriers to burn prevention and our strategy to reduce burning was originally published in *Environmental Management* (Davis & Garb, 2020).

airborne fractions and in the dust that settles around the burn site, later to spread by air, water, and percolating down the soil profile over time. Subsequent exposures can be from breathing, ingesting, or contact with contaminated media dispersed from burn sites, such as resuspended dust, water, and soil, and from local animal products, and produce in contact with these. The bioaccumulation of heavy metals and flame retardants throughout the food chain exacerbates exposure by creating long-lasting effects in the ecosystem and human body, even after initial exposures have subsided (Bellinger et al., 1992, 1994; Ciesielski et al., 2012).

The smoke and smell of e-waste burning have become a backdrop of daily life in the West Line villages. During periods of intensive burning, families that live near active burn sites no longer sleep on their roofs in the summer, for example, for fear of exposure to hazardous smoke, and do their best to seal the edges of their windows and doors with a damp towel to prevent smoke from entering the house. Yet, despite the severe health and environmental consequences in the West Line villages, e-waste burning persists.

Muhammed is one of only a dozen or so "professional" burners responsible for the bulk of cable burning – youngsters whose primary income is offering their services to scrapyards or purchasing cables to extract copper they sell themselves. But, the majority of the West Line population is either directly or indirectly dependent on income from e-waste recycling, which has long hampered the translation of mounting community distress at burning into any meaningful change in practices. Over the years, burning has become so routinized and regularized that community members rarely report burn events and police rarely respond to such reports. Even were reports to be made, it would require a nimble apparatus to apprehend the culprits when a burn is typically over in thirty minutes. Police tire of arriving at the scene to find only a pile of smoldering ash, and citizens tire of having no outcome from their report, and the system had settled into stasis. Several residents reported that the only way they had found to summon police intervention was by having multiple people call them from different locations simultaneously, which would signal a burn event large enough to remain active by the time police arrive. One sustained attempt to prompt action was undertaken at the end of 2011 by a dozen Idhna residents who for two months systematically documented the burn events they reported to the police and the resultant response (or, more typically, lack of response). They sent the detailed spreadsheet with an appeal to the Palestinian Prime Minister's Office in the hope of sparking top-down action, but none was forthcoming.

If enforcement response is faltering within the built-up areas under direct Palestinian police jurisdiction, it is further hampered when burns occur in the areas surrounding these, designated as Area C in the Oslo Agreement. To enter Area C, Palestinian police must first request permission from the Israeli military via the Palestinian District Coordination Office (DCO), which communicates with its Israeli counterpart. The process can take up to forty-eight hours to obtain an answer, which is usually a denial of permission, which would anyway be irrelevant for apprehending burners. Burners clearly know and utilize this fragmentation of governance, so that Area C, especially close to the Israeli separation wall, has become a preferred site for burning. In fact, in some cases when Palestinian police did, nonetheless, approach the areas close to the separation wall, burners explicitly drew upon the deterrent effect of Israeli presence by shaking the motion-sensored chain-linked fence to summon a rapid Israeli military response. In most cases, this would cause the Palestinian police to retreat to their jurisdictional zone in Area B, affording the burners extra time to finish their work. If Israeli military were to arrive at the burn site, unlike Palestinian police they would not confiscate the copper, but would be more likely to fire shots in the distance to scare them away. This offers a more dangerous encounter, but one that permits them to keep their valuable copper.

Close social relations to and even corruption within the local Palestinian police further erode the governance of burning. Locals make appeals to spare those caught, and in exceptional circumstances when police arrest e-waste burners, they are typically charged a small fine between ₪100–1,000 ($28–280) and spend one to three days in jail, or sign a pledge acknowledging that repeat offenses will result in harsher penalties. These punishments are not severe deterrents. Further inducements can be made. For example, one Beit Awwa scrapyard owner told us of his agreement to install an air conditioner in a police officer's home in exchange for hassle-free burning. And some locations where cable burners pay to burn, such as Muhammed's preferred burn site at the outskirts of Deir Samet, seem to be protected from interference – a site that continually burns several tonnes of cables a day would seem to be a viable enforcement task for local police, but they never arrive.

The municipal governments in the West Line villages have considerable potential as agents for regulation, but also liabilities, both stemming from their proximity to the sites and their embedding in a tight-knit community they are policing. For years, municipalities have received complaints of e-waste burning from their citizens, and at times municipal employees have been tasked with enforcement. Since they know the terrain and the locals, they are well-positioned to arrive rapidly at active burn events to identify from afar who is doing the burning and to show considerable nuance in the kind of deterrents brought to play. At the same time, these same ties encumber them. We have witnessed burn events become polarized by clan divisions so that the family of the mayor, the burn response team member, and the apprehended burners all become enmeshed in resentments, appeals, and negotiations surrounding extenuating circumstances and penalties. Thus, the local mayor can take poverty or the recent job loss of the burner into account, releasing confiscated copper in exchange for a written commitment to abstain from future burning, but at some point, this mediation can erode the deterrence of punishment. There is the possibility for the interplay between levels of the governance hierarchy. For example, in the southern villages where there is no police force, the municipal health unit is limited to issuing a fine, but it can then call upon the regional police to arrest the guilty party for nonpayment.

Below the municipal level, there has been a series of sporadic efforts against burning by communities or community sectors. Frustrated with the lack of official response, over the years farmers in the agricultural areas between the villages and the separation wall under heavy burn pressure have grouped together to protect their land. They hired heavy machinery to block off access roads to key burn sites with large boulders, organized regular patrols and volunteer responses with their own vehicles, and in the harvest season of 2020, several clans in Idhna joined together and formed a roster of volunteers to camp out in their fields at nights to ward off burners, declaring their presence and intentions on widely read Facebook forums. Community members have also used Facebook to express dissatisfaction and shame those who engage in burning, often stopping short of naming them personally. In 2019 and 2020, there were protests by schoolchildren and parents sparked by a series of children's cancer deaths. These protests increasingly take on the trappings of civil society efforts worldwide – chants, banners, and petitions. We have also witnessed physical confrontations related to burns – and even between members of the same extended family – particularly when smoke is close to residential areas or when infants and small children are affected. At best, such confrontations push burners to temporarily relocate their activities until tensions subside.

Over the years, there have been repeated appeals to reach out beyond the local level to various regional- and national-level authorities, and even to Israeli authorities. In 2011, for example, farmers complained to the Palestinian Ministry of Agriculture (MoA) that e-waste burning was harming the quality and quantity of their agriculture and livestock production. The MoA initiated an independent burn policing initiative that involved handing out business cards throughout the West Line villages, primarily targeting farmers, with a request to report burn events so that MoA employees could respond and apprehend burners. Yet, similar to the attempts of police enforcement, MoA employees could not reach the sites in time to find burners still present. Though based in Dura, only a fifteen-minute drive from the West Line villages, burners had fled by the time the Ministry team arrived, so that after six months of operation the initiative had not apprehended a single burner and was dissolved. The other agency whose responsibility includes the question of waste burning is the Palestinian Environmental Quality Authority (EQA). The EQA had been approached for years by the community, with little response. Active responses to burn events were hardly an option as their closest EQA office is the Governorate regional headquarters in Hebron, over a thirty-minute drive away, but the municipalities and police requested their assistance in issuing fines to apprehended burners. Here, too, the influence of appeals to the EQA by the family heads of those charged with offenses has seemed to dilute any penalties to an ineffectual minimum.

6.2 Burn Reduction as Part of an Overall E-Waste Strategy

It is against this background of strong sentiment against burning, but equally strong inertia and countervailing barriers, that we attempted to work with the West Line villages toward an intervention that would implement local enforcement as part of a more general strategy. That is, rather than responding to burning as a stand-alone phenomenon to be stamped out, we considered it as an undesirable end point in the recycling industry that offered necessary employment options, and our efforts predicated enforcement on a broader shift to technical and economic alternatives to this hazardous mode of metal extraction. Our efforts benefited from our vantage point of concerned outsiders working to reconcile the hazard and opportunities of e-waste recycling, with the support of the Swedish International Development Agency (Sida) for developing and testing a holistic package of community-based measures. The remainder of this chapter presents the history, rationale, and outcomes of our dual approach, which coordinates opportunities and incentives for change (carrots) in tandem with deterrents (sticks), and considers the applicability of this template to other e-waste hubs in the Global South.

6.2.1 The Carrot: A Cable Burning Alternative

The open burning of cables is simple, rapid, and inexpensive compared to the more environmentally sound methods of copper extraction, which use grinding machinery requiring a large initial investment and considerable knowledge and sophistication to choose, import, operate, and maintain. Some pioneering scrapyard owners have sought some intermediate between the almost free but harmful technology of burning and the insurmountably high investment of a grinding facility. For example, in 2013 one scrapyard owner in Deir Samit inspired perhaps by the traditional water pipe (*nargila*) used for tobacco, constructed a water-fed chamber to filter smoke as it passes from a sealed burning chamber to the chimney. Unfortunately, different designs and experimentations with this homemade incinerator have yet to improve emissions. Strategies to avoid burning altogether can be seen in over twenty large scrapyards that have invested in a $1,000–2,000 cable-stripping machine (see Figure 6.2). These can effectively strip or peel the plastic from thick cables, but cannot compete with open burning as they cannot separate plastic from thin cables, are inconvenient to operate, and have a small capacity taking fifteen hours to process a 100-kg load that could be burned in just thirty minutes. At the high end of such intermediate experiments was the effort of one innovative scrapyard owner who invested $40,000 in a low-capacity copper grinding machine from China able to grind 200–400 kg of cables daily. However, he quickly discovered that he was unable to receive the necessary electrical amperage from the municipality and that a direct high-amperage connection to the Israeli grid (Palestine's electricity

Figure 6.2 A cable-stripping machine found in the West Line villages that can remove the plastic coating from thick copper cables, a time-consuming process compared to burning.

supplier) would cost an additional $60,000. The owner was forced to power his grinding machine with a gasoline-powered generator, and after a month of operation, the high cost of fuel caused him to revert to burning cables.

This abortive attempt with a small cable-grinding machine contained an important lesson. During its brief period of operation, scrapyards from Deir Samit and Idhna began bringing their cables to the new facility, offering to pay for grinding services. Given the machine's small capacity, barely sufficing his own loads, and the high operational costs, the owner had to turn them away. But this pointed to the potential for a cable-grinding facility's viability if the problem of electric supply could be overcome and the scale was large enough to match the supply in the hub. Thus, at a fairly early stage in our efforts in the West Line villages in 2012, we developed an idea with scrapyard owners for a nonprofit cable-grinding social enterprise. Through semi-structured interviews with fifteen scrapyard owners, we explored a creative, viable business model premised on valorizing the plastic outputs and an assessment of scrapyard owners' willingness to pay to transition from burning to grinding (Davis & Garb, 2015). The scrapyard owners gave several reasons for their interest in such a scheme. To begin, grinding produced a higher-grade and higher-priced copper output than burning: A or AA grade, selling at 90–95 percent of the London Metal Exchange (LME) price, compared with burnt "dirty copper," classified as B grade, which sold at 85 percent. There was also the hope that instead of going up in smoke, the plastic retained in the grinding process might have a value if sold to a refinery serving plastic recyclers. In addition, they anticipated that part of the cost of grinding would be offset because it would substitute the cost of burning, which was typically about $100 for two of their employees to burn one tonne of cables at one of the informally "regulated" sites. Finally, they expressed an unquantified but genuine interest in moving away from the individual and collective hazards of burning.

Based on this assessment, we (Davis and Garb) wrote a detailed proposal for a community grinding facility to be funded either as an outright grant or a long-term loan paid back from the eventual proceeds. We approached an international development organization, which showed considerable interest and tried to raise the necessary funding, but ran afoul of elite capture of a powerful political actor who sought to "remove the rubbish people" (i.e., our community partners) from the proceeding and to instead locate the facility in the nearby town of Dura, some twenty kilometers from the West Line hub. The lack of progress with the community cooperative model was a blessing in disguise. While the cable-grinding business model seemed promising on paper, as we learned several years later, there are real limits to basing development projects on asking community members how they will act in certain hypothetical situations – especially when they may feel an implicit pressure to offer favorable responses to an external researcher. Moreover, not long after in 2013, a private entrepreneur appeared on the scene with their own initial funding and a great deal of technical and business savvy to establish a similar facility on a private commercial basis, called Safa Recycling.[3]

Safa Recycling was forging a business based on two revenue streams. One was the direct sourcing of high-quality, thick cables from Israel,

[3] "Safa" is an Arabic word that translates to "pure" in English, which was adopted as a name to brand the clean copper extracted from their facility.

which have a high copper content and few unwanted plastic attachments. The other was processing heterogeneous batches of cables that local West Line scrapyards had removed from scrapped cars and waste. Here, the profit was considerably lower because of the labor required to sort and trim the cables before grinding and the higher wear and tear on the machines, but this stream provided steady activity even during periods of temporary loss of access to Israeli markets. The grinder's main output was high-quality copper sold on the international market, with a side product of granulated plastic pellets from the insulation, which was stored locally with plans to find a buyer or else an investor who would allow the purchase of machinery for in-house plastic recycling.

With its existing facility and a great deal of hard-won expertise, Safa Recycling had the potential to be a critical leverage point for systemic change. With some adjustment, it might offer a private-sector pathway to some of the goals we had anticipated in forwarding the communally owned grinding facility. Their grinding facility offered a concrete way to reduce the level of burning and begin redirecting the overall system, potentially in a sustainable, subsidy-free fashion. Not only did the business reduce burning, but it also tilted discourses and expectations in a new direction. Locals might be more willing to report or pressure burners once they knew that an alternative mode of copper extraction was available, while informal businesses would have a live example of the benefits of a transition to legal and clean operation – in particular, the ability to import cables directly without smuggling and fear of confiscation.

We began negotiations with Safa Recycling to determine the costs and protocols to rent their facility for a period in which, as part of the Sida project, they would provide subsidized cable-grinding for local scrapyards, gradually transitioning to subsidy-free operation. This process began with an initial six-month free cable-grinding promotion, which opened Safa Recycling to accept cables from local scrapyards (see Figure 6.3). This promotional period enabled us to test the costs and procedural operation of a publicly available grinding facility and test the willingness of burning businesses to transition to a grinding service model. All loads were carefully monitored, showing that during this trial period nearly three tonnes of cables were ground daily, with the West Line village locals reporting an 80 percent reduction in burn events. Remarkably, this shift to grinding occurred without an increase in police enforcement.

Figure 6.3 Posters advertising cable-grinding distributed throughout the West Line villages, targeting scrapyards. Translated to English, the poster reads: "Free grinding of plastic cables offered instead of burning them. The offer is available from 09/01/2017 through 02/15/2018. Through a Swedish Grant, the Safa Recycling Facility allows you to grind your cables free of fees. Go to Safa Recycling to sell or grind your cables without fees, so long as they are acceptable and will not need a big effort to clean. The Safa Recycling address: Wadi Rishi - Idna - Hebron. Phone: 0568448804."

We then shifted to the challenge of transitioning this pilot effort to a long-term, financially sustainable cable-grinding solution. During this period, Safa Recycling had dedicated the entire facility to grinding local cables, with the Sida support compensating for this "rental" of their facility. At the same time, while scrapyard owners had initially declared their willingness to pay $100/tonne for grinding as a service, in practice, when we began to charge for grinding, the lure of "free" burning was too much and many reverted to this practice.[4] To begin reducing subsidies, we needed to lower the cost of grinding and raise the cost of burning so that the business could operate on a purely commercial basis.

[4] This revelation speaks to a broader literature on the fallacy of "willingness to pay" feasibility studies to rationalize development projects, such as the provision of clean drinking water with continuous costs (Abramson et al., 2011).

We patched together several means to this end. One was a continuation of the facility rental/subsidy arrangement for an additional period, but with an upfront payment of around $250,000 as an advance for the rental to be used specifically for the purchase of equipment that would allow the high-quality cables from Israel and the low-quality local cables to be processed at the same time, rather than needing to forgo the former for the latter. The facility could then operate ongoingly in a cross-subsidy mode, with the higher profit stream from Israel allowing the processing of the local cables at close to breakeven. Simultaneously, we helped local municipalities begin the local enforcement efforts that would raise the "cost" of burning. In addition, the owners of Safa Recycling suggested a creative way of bridging the gap between the burning and grinding costs, through proposing that the Palestinian Authority (PA) waive the 16 percent value-added tax (VAT) on the sale of copper extracted from grinding cables. Initial modeling showed that this differential was close to the difference between formal (grinding) and informal (burning) modes of operation.[5] We made a case for such a waiver for a five-year period to the Palestinian Prime Minister's Office and subsequently the Ministry of Finance. Our argument was based on pointing to precedents internationally in which governments granted VAT waivers for environmentally or socially virtuous enterprises, and on the argument that the waiver was, in fact, not going to decrease government revenues, as no taxes were being paid on the cables flowing through the informal system. Thus, granting a five-year transition period for redirection of these untaxed cables to formal businesses would allow the infrastructures and practices to stabilize to the point where they could begin to contribute taxes in due time.

[5] The rationale for a VAT waiver works as follows. Cables extracted from dismantling e-waste (i.e., "thin cables") typically cost around $350 per tonne to prepare and grind, producing roughly 350 kg of copper and 650 kg of plastic pellets. With current international copper prices hovering around $5.20 per kg, copper sales from grinding a tonne of cables would generate $1,820 ($5.2 × 350 kg). A 16 percent VAT waiver applied to the copper sales would save Safa Recycling $291 ($1,820 × 16 percent), which would reduce the cost of grinding by 83 percent. Since copper extracted from burning cables is sold informally, without any tax paid to the Palestinian Authority, the VAT waived on copper sold from grinding these same cables would not constitute a net revenue loss to the Palestinian Authority.

There was one more component to make the grinding option more competitive that proved to be just as elusive: capturing the potential value of the plastic, which in the grinding facility was recovered as a potential source of value rather than burned to become a severe hazard. The tonnes of recovered plastic were building up in huge bags in and around Safa Recycling's facility, but finding a taker that was both environmentally responsible and profitable was difficult. We were hesitant to send this plastic stream that carried burdens of lead and flame retardants to recyclers that would mix it with the more general plastic waste stream, but the more stringent and environmentally responsible operations did not offer a reasonable business model – a dilemma that we learned is a global one that has yet to be resolved.[6]

While the main focus of this chapter is on extraction burning, we should mention in passing the disposal burning of the by-products of the e-waste industry as well as of municipal solid waste, which is all too often burned. Such burns have different dynamics and solutions than extraction burning. In health terms, these materials – especially scrap dismantling by-products – contain plastics, flame retardants, foams, and other chemicals in glues or paints that are also hazardous, and their fate is entwined with extraction burning in other ways. For example, scrapyards and households that do not have formal avenues for disposal might dump their waste in sites of frequent e-waste burning, with the expectation that it will be burned at the same time as extraction burns – expanding the size of the burns and making our sampling of burn sites a doubly messy business. If such burning persists after extraction burning has been reduced, it will also hamper the burn site rehabilitation efforts discussed in the next chapter.

On occasion, citizens and enforcement teams were triggered by smoke and smells issuing from a disposal burn, making it less possible to cleanly evaluate the impact of measures designed to reduce extraction burning. Recently (2020), for example, Israeli neighborhoods to

[6] The plastic pellets comprise a mix of rubber, polyvinyl chloride (PVC), polyethylene (PE), and polypropylene (PP) plastic types. Local factories that can purchase and repurpose these plastic outputs require this mixture to be separated into a homogenous mix based on their melting temperature, which is a separation process Safa Recycling has yet to resolve. Furthermore, the price of plastics is closely tied to that of oil, which can make any solution financially vulnerable to global commodity markets. As the price of oil increases, the plastic outputs will sell for a higher price that could raise the appeal of cable-grinding over burning, although the business is at risk of downturns as well.

Figure 6.4 This landfill, located south of Idhna, served as a convenient site for scrapyards to dispose of non-valuable materials until it caught fire in March 2015. The smoke emitted from this fire was much more extensive than burning cables, and required three days for firefighters to extinguish it.

the west of Idhna urged the army to respond to burns that appeared to be a major cable burn, but which we knew from locals to be from the disposal of a large load of plastic and upholstery from the car recycling industry, or, in another case, from an accidental fire of discarded waste. In fact, some of the largest and most iconic burn events in the West Line villages resulted from homemade landfills catching fire and burning months' worth of dumped e-waste discards. One infamous burn event in 2015 required three days of sustained effort from firefighters to put out the fire, and smoke could be seen by *Alte Zachen* collectors working in Beer Sheva over forty km away (see Figure 6.4). Because they are difficult to distinguish from one another, burn response and enforcement teams must be prepared to handle both kinds of burns, though penalties are more difficult for disposal burns, where those responsible do not need to wait half an hour for their valuable metals, but can simply ignite a pile and flee.

Since waste disposal burning is always a tempting alternative when municipal collection systems are inadequate or landfilling too expensive, it is a problem in areas of marginalization and patchy governance all over the West Bank, and often in the Arab municipalities in the north or Negev regions in Israel. Until 2016, the closest landfill to the

West Line villages was in Yatta, a thirty-minute drive from most e-waste dismantlers. Considering the cost of gas, labor, and the tipping fee, and the fact that many of them did not have licensed vehicles that could go beyond the villages, it was only the largest scrapyards that generated sufficient quantities of discards and could afford to use these formal facilities. In 2016, as part of the Hebron Governorate's investment in solid waste management and the opening of a large adequately sealed modern landfill in the Bethlehem area primarily funded by the World Bank, an additional waste transfer station was established in Tarqumiya. This waste transfer station was located just north of the West Line villages and a ten-minute drive from the heart of the e-waste hub. The Idhna municipal government subsidized the tipping fee for scrapyards to encourage their use of the waste transfer station. This increased usage among the medium-to-large scrapyards, but still did not serve the needs of smaller backyard scrapyards, especially those located further away in Deir Samet and Beit Awwa. Through the Sida project, we attempted to assist these local municipalities in establishing their own local transfer station. However, it proved difficult to find suitable land in the municipal boundaries, while land in the surrounding Area C, and under Israeli civil jurisdiction, would have required a lengthy environmental assessment and planning approval process.

6.2.2 *The Stick: Community Environmental Policing*

Prior to our pilot project that demonstrated promising results transitioning from burning to grinding cables with incentives alone, the stick of enforcement had been the preferred management instrument to reduce e-waste burning used by the Civil Administration and Palestinian government branches. Israeli and Palestinian environmental authorities expressed a real hesitancy to engage with solutions outside of strict enforcement. For example, several times in making requests to the Civil Administration for approvals that would facilitate solid waste disposal, such as the transfer station described above, or the transition of e-waste businesses to environmentally sustainable operation (discussed further in Chapter 9), we were told: Show us that the burning goes down for half a year and then we will have the confidence to do the things you are suggesting. We argued that things could proceed the other way: The measures we were requesting would reduce burning. In the end, it took many months of futile efforts by the

Israeli authorities to try to reduce burning through unilateral force, on the one hand, combined with our own demonstration, described above, that the provision of alternatives did drastically reduce burning without any increase in policing, to really command their attention and increase their responsiveness to our "out of the box" requests. This episode strengthened our intuition that while enforcement and inducement are bound together, the latter is a more ethical and effective way to break the chicken-and-egg deadlock of such situations. Indeed, throughout our Sida-sponsored project, and working with the private sector and community group partners, we did not have enforcement in our toolkit. We did, however, come to appreciate its importance as part of a two-pronged solution to reduce e-waste burning and were able to lay the groundwork for an eventual increase in enforcement against burning as a later adjunct to our efforts to first provide more appealing options and pathways.

Our main effort in this direction was a community-based alert and response initiative that simultaneously allowed us to (1) develop and test whether it was possible to detect, locate, and reach burns rapidly enough for deterrence to occur; (2) more deeply understand the patterns and locations of burning; and (3) inform burners of the new subsidized grinding facility and invite them to use this. This initiative consisted of recruiting and training concerned citizens to systematically report burn events to a 24/7 "burn alert coordinator," who would then call one of two teams of "burn responders" to arrive at active burn sites, interview burners, and issue them a voucher for free grinding at the Safa Recycling facility.

Our system was forged on several constraints. First, as mentioned, burn events might go from the first smoke to the departure of the burners within thirty minutes, providing a brief window of time for a response unit to arrive. Second, while burns emit massive plumes of smoke that can be seen for kilometers even if started in deep valleys, the landscape is complex with literally hundreds of potential burn locations, making it difficult to locate and arrive at the correct site. Third, people's ability to contact the burn alert coordinator in a rural context was quite heterogeneous, with patchy Internet or smartphone availability, so that messaging needed multiple channels and modalities for report. Fourth, anonymity is vital; villagers do not want to risk being perceived as an "informant" reporting environmental crimes against neighbors and in some cases family members. Finally, our budget for a community environmental policing campaign was limited to at most $10,000 for a three-month period.

These constraints precluded a large team or reliance on dedicated high-technology solutions such as hidden cameras at burn sites, automated drones with heat sensors, or Global Positioning System (GPS)-enabled smartphone reporting applications. Rather, we used word of mouth and snowballing outreach to widely dispersed local community members living close to or with good visibility of the frequently used burn areas, and with demonstrated concern about burning. In this way, we rapidly identified fifty-five volunteer reporters, who were registered in a confidential database that recorded their contact information, GPS coordinates, and the burn sites located within eyesight of their homes. This generated an online dynamic map (based on Google Maps) accessible only to a "burn alert coordinator," who received burn reports, determined their location, and then immediately directed one of two "burn responder" teams to arrive at the site. The burn alert coordinator served as a critical buffer, so that burn responders were not aware of who had reported burn events when they contacted burners, thus protecting the anonymity of reporters.

This extensive and geographically dispersed registry of volunteer reporters, coupled with a prior training with them, enabled our team to rapidly identify the location of reported burn events. Our team sat with each volunteer in advance to work out a simple but highly tailored procedure in which they could report burns in a way that allowed rapid and precise locating of these by the coordinator. With no systematic named street system to speak of, and most people not able to orient with respect to the compass cardinal directions, we found it simplest to work based on a culturally familiar and reliable orienting device, namely reference to the direction of Mecca, where Muslims face for daily prayers, which is roughly south of the West Line villages. In addition, for each reporter, we established some of the key landmarks familiar to them in their own area. This enabled us to determine the location of burn events in relation to the home of the reporter. And if a burn reporter could not pinpoint the exact location of a burn event that was obscured or was several kilometers away from their location, the burn alert coordinator would contact other reporters who live closer to the burn event to confirm the exact location. Reporters were given several options to report burns: phone calls (both Israeli and Palestinian numbers), SMS, email, private Facebook messages, and

WhatsApp. Since most volunteer reporters were on costly pay-by-the-minute cell phone plans, the understanding was that the burn alert coordinator would hang up upon receiving a call and then immediately call back to eliminate any cost to the volunteer reporter.

Regardless of the platform used for notification, the burn alert coordinator would clarify the location of the burn and when it began, and then pass this information on to one of two burn responder teams. These were strategically located in the villages' northern and southern ends and remained on call 24/7 to respond to burn alerts and rapidly arrive at burn sites. For each reported burn event, we kept a record of the location of both the burn responder and the burn event, along with the time notified and time of arrival at the burn. This gave us a better sense of the distance burn responders could cover and still reach burn events while still active. Once at the site, the response team interviewed burners, recording their age, the material burnt, compensation for burning, rationale of burn site selection, and how often they burned, and informed them of the free cable-grinding promotion.

In this way, our community-based alert and response initiative achieved an efficient and flexible system that was continually available to respond to burn events. The total cost was three salaried full-time workers (one burn alert coordinator and two burn responders with cars), and one dual-SIM smartphone for the coordinator with Israeli and Palestinian SIM cards. The site and volunteer locations were managed on Google Maps and the burn log on a Google Docs spreadsheet, with the phone connected to a dedicated WhatsApp account and Facebook page to receive private "burn alert messages." The burn alert system operated from December 18, 2016, to March 31, 2017, and geared up incrementally: first the volunteer reporters and a burn alert coordinator for the first month; then, from January 25 to March 31, 2017, the first burn responder in the southern region of the villages was active, and from February 20 to March 31, 2017, a second burn responder was hired to cover the northern region of the villages.

This low-cost and low-tech system proved remarkably robust and successful. During the 104 days of operation, a total of 412 burns were reported (averaging four a day), though we expect that other burn events were not reported, especially those that occurred when reporters were sleeping. Of the 412 reported burns, 115 events resulted in the dispatch of burn responders, whereas the other 297 burn events were reported after the burn event had already finished. Burn responders were able to reach the site while burners were there in about 85 percent of the cases they were dispatched. The average straight-line distance traveled to each burn site was 2.15 km, which took, on average, twenty-five minutes. While this may seem like a long time, it should be noted that the road network in the West Line villages is convoluted and unpaved in many sections, particularly in the more rural areas where burns tend to occur, and burn responders were not always able to disengage immediately from what they were doing and respond. The alert-to-arrival time for the southern village of Beit Awwa was much shorter, an impressive average of seven minutes. Of the 115 burn events that burn responders documented, 80 percent were extraction burns and 20 percent disposal burns. The ratio differs between the southern and northern villages, with more disposal burning occurring in the former, which are far from the waste transfer station of Tarqumiya through which municipal waste is taken to the large regional landfill near Bethlehem. The distribution of sites and burners in these 115 burn events was skewed: Forty-three unique burn sites were visited throughout the West Line villages, with the overwhelming majority located in Area C, with only thirty-three individuals responsible for all reported burns, of whom seven accounted for over half of all burn events. All the burners were male, with an average age of thirty years.

Word of the free grinding option spread to burners rapidly, but our interviews then revealed a surprising, disconcerting finding: They preferred burning even when the free cable-grinding promotion was available. Several reasons for this emerged, sharpening our sense of additional measures that would be needed in the future to overcome this hesitation. Unexpected economic considerations were at play. Even though ground cables produce a higher grade of copper that is priced at 5–10 percent higher than burnt "dirty" cables, burners were under the impression that ash and dirt mixed with the copper during burning increased the weight of the load and more than offset this loss. The timeliness and flexibility of transactions were also important for these small-scale informal operators. With grinding in someone else's facility, they would need to accumulate some minimal amount of cables to make the transport to the facility feasible and wait several days for their load to be processed. With burning, they can buy cables, process a small load immediately, and sell the extracted copper on the same day. This speed has the advantage of being immediately able to use these funds to buy the next load, and it minimizes the risk of losses of fluctuating copper prices. (A larger facility can wait and sell once copper prices rise again after a dip.)

We initially thought that burners would be eager to outsource copper extraction from the dirty hazardous burning process they did themselves to grinding at an external facility. But we did not realize the degree to which they discounted the dangers, nor the advantages to them of using the familiar in-house processing method they had used for years. When cables are brought to the grinding facility, the client is presented with two options: selling the cables for a negotiated price based on the estimated amount of copper they contain, or selling the copper after it has been extracted. While Safa Recycling attempted to offer copper prices that were higher than the going rate elsewhere, the burners were mistrustful of both options. They feared that the copper content would be higher than estimated if they sold the cables on the spot, or that not all of it would be presented if they left the cables and returned some days later after the cleaning and grinding were completed. Thus, having the entire process under their own control outweighed the risk to themselves and others.

These realizations sharpened our sense that while the grinding option and enforcement were important pillars for moving away from burning, additional systemic changes would be needed to overcome the inertia and incentive structures that made burning preferable. A holistic approach would need to include measures to ensure the lone grinding operation does not exert undue power and to reduce the information asymmetries and capital buffering problems that made burners hesitant or suspicious regarding the grinding option. And, beyond that, we would need to boost the incentives and pathways away from burning for the small cohort of "professional burners" who had come of age working as burners. For example, our Sida project prioritized employing them in the cleanup efforts, which would assure a couple of years of transitional income, and we lobbied the Civil Administration to give them priority in the allocation of permits to enter Israel for work. Also important would be to find solutions for disposal burning, which, while not directly related to the dominant cable burning phenomenon, was sufficiently entwined with it to be a necessary part of a solution package.

6.3 Concluding Remarks

The use of open-air burning as a deleterious, yet effective way to extract metals from e-waste and, in particular, cables is common in the Global South, reaching oppressive proportions in and around the world's e-waste hubs. In this chapter, we have seen some of the incentives for and dangers of these practices and detailed how we piloted a two-pronged approach to reduce burns – mechanical grinding and local enforcement – which clearly demonstrated the viability and effectiveness of both.

Various political impediments undermined the transition of this pilot into a longer-term project, but the basic ideas were taken up by local actors. The EQA was unwilling to proceed with the larger project that would have expanded and institutionalized the transition of the recycling sector to clean operation that the cable facility had pioneered, and with enforcement done at the local level where rapid and community-sensitive response was possible. Instead, it insisted that all e-waste coming from Israel must be stopped, rather than helping the productive informal Palestinian recycling sector pivot toward clean and formal operation. Similarly, it insisted that any environmental policing was in its jurisdiction alone, and that the salaried positions and patrol vehicles allocated to this end must reside in the EQA central offices, whether in Ramallah or Hebron.

The Safa Recycling facility we helped institutionalize expanded to grind both local and higher-quality Israeli cables in parallel, as we had hoped, and now grinds around ten tonnes a day, fairly evenly balanced between these two sources. Thus, while burning continues with little local or PA enforcement, the facility still prevents at least half of the potential amount that might be burned, and this is on an entirely commercial, subsidy-free basis. After many attempts, our request for a VAT waiver for formal cleanly operating recycling facilities was considered for approval; however, the authorities mandated that it would apply only to facilities that operate under the pretense of processing only indigenous Palestinian waste, not e-waste from Israel that has been declared as illegal regardless of whether its destination is an authorized facility. Since 90 percent of the waste processed in the West Line originated in Israel, a formal facility such as Safa Recycling whose business model depends on coordination of both streams will not be eligible.

After the successful piloting of our alert and response system, it was taken over by the Idhna and Beit Awwa municipalities' health departments. They could do more than just issue grinding vouchers and interview burners; they confiscated cables, gave fines, and pursued burners. In fact, in an unprecedented move, based on the drastic reduction in burns during the pilot project, we were able to facilitate an agreement between these municipal teams and the Civil Administration that allowed them to gain real-time approval to enter Area C in pursuit of an ongoing burn.

However, the municipal burn response units used the municipality vehicles, which resulted in more overtime for the workers, whose standard hours are only until three pm, and entailed more costs for fuel and maintenance of the vehicles. The Sida project contributed toward these additional small expenses for some months, but after this, the PA agencies clamped down and these activities were stopped. The hotline arrangement we had brokered allowing the Palestinian municipal health unit to access Area C in pursuit of a burning event was also stopped. Though this arrangement had drastically expanded the autonomy and range of Palestinian governance in Area C, it sidelined the central Palestinian authorities, who said that all discussions with Israel must go through the DCO of the PA based in Hebron, which made the arrangement too cumbersome to be relevant. While the model we forged and tested came to an end because of these political pressures, we demonstrated it to be highly successful. It has much to offer as a template that might be taken up in other places, or in the West Line villages, if and when circumstances change.

7 *The Toxic Elephant in the Room*

The West Line burn sites are large enough to be visible from space. Take Wad Gura, one of the oldest, largest, and most iconic of these sites, located in a deeply incised valley just south of the road between Deir Samet and Dura (see Figure 7.1), one of the upper reaches of the ephemeral Lachish stream that drains the Hebron mountains to the Mediterranean. Even in Google satellite imagery (31°31′03.3″N 34°59′22.4″E), one can discern almost a kilometer of black stain along the bottom of a valley, marking close to two decades of accumulated toxic ash from the daily burning of e-waste and other waste materials. In addition to the loads of cables, plastics, and foams generated by the e-waste industry, it has become a convenient place for residents to avoid hassle and tipping fees when they dispose of household waste, tires, construction materials, and even livestock carcasses.

The crisp scar in the aerial view cannot convey the visceral immersive impact of a visit to this toxic wasteland. One sets out from a main intersection of the roads from Al Kum to the north, Deir Samet to the south, and the Hebron outskirts of Dura to the east, along a narrow farm road lined with olive trees. But after a few hundred meters, the traces of burning intensify until one is in a barren valley floor where half a dozen burn sites and years of truck traffic between these have coalesced into a black dusty carpet that sticks to your tires and feet and rises to your eyes, nose, and lungs as you walk. Any patch you examine is full of the small and even microscopic remnants of the many tonnes of e-waste that have ended up here: hair-like copper strands from cables, chunky shards of thick lead-impregnated glass from television and computer cathode ray tubes (CRTs), bits of plastic, foam, and rubber. Even when there is no active burn, the penetrating chemical smell is oppressive – the West Line smell.

Wad Gura is just one of many notorious burn sites in the area. Our systematic remote sensing analysis of those active up until 2011 showed hundreds in the area, with roughly thirty of them heavily used over an

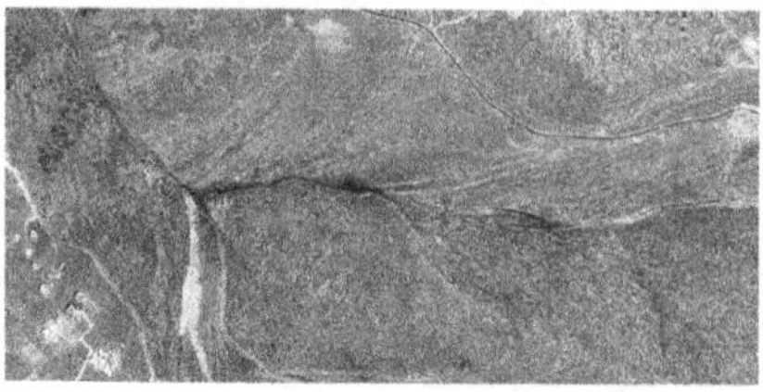

Figure 7.1 Aerial imagery of Wad Gura. At the top right of the image is the road connecting Deir Samet to Dura. Below, at the valley bottom, are three of the main sites where e-waste is burned, connected by a dirt road. The bottom left of the image shows Palestinian homes in close proximity.

extended period (Davis & Garb, 2018a). Bidoo Cave (31°32′23.5″N 34°57′00.2″E), for example, is of considerable heritage value, an archaeological site similar to the Bell Caves located twelve km northwest in Beit Guvrin-Maresha National Park, which is designated as a UNESCO World Heritage Site in Israel (see Figure 7.2). The burn sites around Khalet Saad (31°32′52.5″N 34°56′47.7″E), near the separation wall, are surrounded by olive trees, agricultural fields, and the grazing areas of livestock, and burns here often cause consternation to the residents of the Israeli neighborhood of Eliav, a short distance away on the other side of the separation wall. Another infamous longstanding burn is adjacent to a quarry, east of Idhna, on the hilltop near the small village of Suba (31°32′45.0″N 35°00′18.5″E). We have measured levels of lead that exceed those allowable for agricultural lands even a kilometer downstream in the valley of this site, limiting what crops can be grown there, and there is almost no house in Suba without cancer or congenital disability.

In Wad Gura and other heavily used burn sites, the accumulation of ash from frequent e-waste burning has become oppressive enough to bother even the professional burners, making it difficult to drive trucks and hard to breathe, especially if there is a wind. Periodically they collect funds among themselves to hire a bulldozer operator to push loose ash to the road's edges and dump a thin layer of topsoil on the surface. This reduces some of the immediate hazard and inconvenience and perhaps the wind-borne spread of contaminants, but does little to reduce the downward seepage of contaminants. In fact, this cosmetic covering of burn sites in the West Line is common, when the landowner of a burn site wants to "clean" their land to increase the sale

Figure 7.2 The Bidoo Cave. This archaeological site is now used as an e-waste burn site as it conveniently hides much of the smoky process. The small rectangular alcoves on the back wall indicate where ancient residents raised pigeons for both consumption and ritual sacrifice.

value or repurpose it to build a house or school. But by masking the ash, the site identification and remediation we hope for only becomes more difficult.

West Line inhabitants' concerns about contamination are overwhelmingly supported by a vast and growing scientific literature. Similarly, there has been an explosion of literature, policy consideration, and regulation on the e-waste problem. Yet, there is little intersection between the contaminant-focused scientific literature and the policy-focused e-waste management literature.[1] Though the scope and urgency of the e-waste contamination legacy are fearfully clear, the discussion of e-waste solutions is forward-looking, about stopping future contamination without considering the decades of this already on the ground. Thus, the massively described hazards central to scientific literature barely enter into the discussions of policy measures, other than as a spur to action. We observed the same disconnection in our local context. While both Israeli and Palestinian actors and commentators eagerly foregrounded the environmental and health effects of improper e-waste treatment as urgently motivating new policies and practices, it was only after considerable difficulty that we were able to place the need for cleaning up the existing sources of contamination from past practices firmly onto the table for discussion, funding, and action.

This chapter shines a light on the contamination legacy within e-waste hubs and contends that remediation, once contamination sources have been significantly reduced, must be a key component of e-waste management policies in the Israel–Palestine context and beyond. We first summarize some of the rapidly expanding international scientific literature on contaminants within e-waste burn sites, on the one hand, and the lag in policy consideration of how their presence and impacts should be remediated, on the other. We then describe our own piloting and development of a comprehensive

[1] Cazabon et al. (2017) offers an exception. This study conducted a Delphi round of interviews with nineteen stakeholders from academia, government, private sector, and civil society to determine the pressing problems arising from e-waste activities at Agbogbloshie as well as suggested solutions to these. Notably, none of the interviewed stakeholders were representatives working in the informal sector. The results of the Delphi interview process produced a ranked list of the fifteen most beneficial and feasible response options. Remediation of contaminated sites was listed as the sixth best response option in terms of both impact and feasibility.

cleanup plan, and how this was stymied by the frustrating politics within the Israel-Palestinian setting.

7.1 The Enduring Harm of E-Waste Burn Sites

In the summer of 2016, we began to conduct systematic analyses on some aspects of the nature, extent, implications, and dispersal/uptake mechanisms of contaminants in e-waste burn sites. For example, we used a combination of extensive measurement using a portable X-ray fluorescence (XRF) spectrometer and more focused laboratory analyses using inductively coupled plasma (ICP) and other analytic techniques to measure contaminant levels in heavily used burn sites and in substrates and surfaces at and around burn sites throughout the West Line, as well as the streams and areas adjacent to the separation wall on the Israeli side. Our emphasis in these measurements was primarily on the heavy metals, which are far cheaper and easier to measure and which probably serve as a reasonable proxy for the presence of high concentrations of a range of organic pollutants. The levels we measured were remarkably high – predictably reaching over 10,000 parts per million of lead in burn site cores and dropping over the first hundred meters or so; however, depending on the circumstances of transport, they were sometimes still measurable at low levels at considerable distances.

Our findings are aligned with the scientific literature on contamination at e-waste processing and burn sites, which has grown enormously over the last decade. Because of the complex and heterogeneous composition of e-waste and the various and often harmful ways through which it is processed or disposed of in e-waste hubs, this literature records a remarkable range of worrying chemicals that are generated and dispersed. The cataloging of what is known about the dispersal, uptake, and impacts of these substances constitutes a massive and ever-growing task, complicated by the fact that despite an emerging core cluster of recurring substances and themes, each e-waste site's chemical composition is specific to its material flow, disposal practices, and geography. For this reason, we provide a detailed review of contaminant concentrations within e-waste burn sites, their dispersal, and transmission pathways, before turning to the question of their remediation.

Heavy metal soil contamination is a significant environmental problem and one that received the earliest attention in the e-waste literature, as the environmental chemistry and impacts of some heavy metals, such as lead, have been known for a long time, and they are also somewhat easier and less expensive to measure. The accumulation of metals in soils can degrade soil quality, reduce crop yield, and lower the yield and quality of agricultural products (Nagajyoti et al., 2010), but this depends on various factors, such as pH level or organic content. A growing body of research, primarily from e-waste hubs in China (e.g., Taizhou, Qingyuan, Guiyu, Longtang, and Fengjiang), has analyzed heavy metal concentrations in soils at and around e-waste processing areas, with the results compared to soil guideline values of different regulatory bodies to estimate local health risks. For example, in Wenling, China, Tang et al. (2010) found elevated levels of mercury, copper, cadmium, lead, and zinc in agricultural soil samples, with mercury the most severe heavy metal pollutant with levels exceeding Chinese environmental quality standards by one to thirty-one times. Wang et al.'s (2015) analysis of heavy metals in soil samples near e-waste dismantling sites in Longtang, China, found that the mean contents of copper, zinc, lead, cadmium, and mercury exceeded their corresponding Dutch acceptable values by 17.9, 3.0, 2.2, 13.8, and 13.6 times, respectively. Li et al. (2011) found that lead, cadmium, copper, and zinc were the most enriched metals found in soil samples from open burning sites in Guiyu, with the metal contents exceeding Chinese environmental quality standards for soils by 13.7, 50, 363, and 35 times, respectively.

Similar findings have been replicated in Ghana (Ackah, et al., 2019; Akortia, 2017), Nigeria (Adeyi & Oyeleke, 2017), Thailand (Damrongsiri et al., 2016), the Philippines (Fujimori & Tukgami, 2014), Vietnam (Oguri et al., 2018), Cameroon (Ouabo et al., 2020), India (Pradhan & Kumar, 2014), and Indonesia (Soetrisno, 2020). Moreover, the effects of pollution can amplify through metal migration, enrichment, and transformation (Zhao et al., 2015). In our area, lead is of key concern, with levels at burn cores twenty to thirty times higher than are allowable on agricultural soils, copper is also extremely elevated, and some sites have very high levels of arsenic. Exposure to high concentrations of heavy metals can have severe health implications. For example, lead can interfere with behavior and learning abilities, copper can result in liver damage, mercury can damage motor skills and cause intelligence disorders, and cadmium can increase the risk of lung cancer and kidney damage (Song & Li, 2014).

In addition to heavy metals, a range of persistent organic pollutants (POPs) are generated by e-waste burning, such as polycyclic aromatic hydrocarbons (PAHs), dioxins and dioxin-like substances, and polychlorinated biphenyls (PCBs), which have all been identified in soil samples from e-waste hubs (Leung et al., 2006; Wong, 2007; Zhang et al., 2012). The abundance of polyvinyl chloride (PVC) as a chlorine source and copper (in wires and cables) coupled with relatively low temperatures during the open burning of e-waste generates complex mixtures of these substances.

Maliszewska-Kordybach (1996) suggested four levels of soil contamination for PAHs: non-contaminated soil (<200 ng/g), weakly contaminated soil (200 to 600 ng/g), contaminated soil (600 to 1,000 ng/g), and heavily contaminated soil (>1,000 ng/g). PAH concentrations are generally higher near incineration sites, whereas the surrounding area is contaminated due to the deposition of fly ash from burning (Yu et al., 2006). Using this classification, Huang et al.'s (2014) analysis of PAHs in Longtang, China, showed that open burning sites (~2,300 ng/g) far exceeded the criteria for heavily contaminated soil, dismantling sites (~680 ng/g) were contaminated, and wastelands (~260 ng/g) were weakly contaminated. PAHs are transported through the soil profiles, including those in agricultural fields, and have been detected at depths of 80 cm (Huang et al., 2014). They are of concern due to their genotoxic, carcinogenic, and teratogenic potential. There are over 100 known PAH compounds, sixteen of which have been identified as "priority pollutants" by both China and the United States Environmental Protection Agency (US EPA), and seven of them are considered probable carcinogens (Cai et al., 2007; Leung et al., 2015). PAHs are also resistant to biodegradation and can bioaccumulate, which introduces a long-term threat to the environment and human health (Huang et al., 2014).

Polychlorinated dibenzo-p-dioxins and dibenzofurans (PCDD/Fs) have similar patterns of generation and dispersal. In Guiyu, Wong et al. (2007) analyzed PCDD/F concentrations from surface soil samples in open burning sites, areas near open burning sites, rice fields, and reservoir areas. Apart from reservoir areas, soils collected at all sampling sites were above the Canadian soil guideline value of PCDD/Fs for agricultural, residential, commercial, and industrial land uses. Moreover, samples from open burning sites exceeded the value that would trigger remediation, as adopted in the United States. Tang et al.'s (2010) analysis of PCBs in agricultural soil in Wenling found that concentrations were almost twice as high as allowable levels in ambient soil according to Dutch, Australian, New Zealand, and Canadian soil guidelines for residential areas. Polychlorinated dibenzo-p-dioxins and dibenzofurans are particularly concerning since they have the potential for bioaccumulation, long half-lives in the environment and biota, and can cause a range of adverse health effects, including reproductive and developmental abnormalities, suppression of the immune system, tumor promotion, and endocrine disruption (Chan et al., 2013; Hedley et al., 2006; van Leeuwen et al., 2000). Flame retardants, especially polybrominated diphenyl ethers (PBDEs), are pervasive in electronics and electrical devices, with especially high levels in older equipment. They can also bioaccumulate and cause thyroid hormone disruption, neurodevelopmental effects, and cancer (McDonald, 2002).

The soils and deposited ash of burn sites are a primary initial receptor of heavy metals and organic contaminants and are heavily researched (Auclair et al., 2020), but they then become a secondary source of transmission onward into the air, food, and water bodies as well as groundwater, and then enter food chains. While heavy metals tend to remain close to the pollution source in open burning sites (Fujimori & Takigami, 2014; Li et al., 2011), they may be transferred to soils and sediments by leaching or carried onward on dust particles (Zhang et al., 2012). Heavy metal contamination through wind and water transportation has affected adjacent areas, including paddy fields and vegetable gardens, and they have accumulated in various plants resulting in ecotoxicological effects (Fu et al., 2008; Jun-hui & Hang, 2009; Luo et al., 2011; Olafisoye et al., 2013; Pradhan & Kumar, 2014).

In contrast to heavy metals, POPs have a significantly wider "halo effect." Li et al. (2011) reported that the transport range of PBDEs and PCBs was forty km, with heavy metals remaining at the original site without transport to the surrounding areas. Zhao et al. (2009) reported that PBDEs could be transported even further from e-waste recycling sites, up to seventy-four km using *Cinnamomum camphora* leaves (an evergreen tree species) as biomonitors. However, despite the limited mobility of metals – especially in an arid environment and when acids are not used in e-waste processing (Friedlander et al., 2019) – our own research with metal concentrations in the West Line area confirms indications in the literature that there may be considerable variability and surprises with the dispersal patterns of metals. For example, Mudge et al. (2019) showed a significant enrichment in metals associated with solder and copper dispersed two kilometers downwind from the source material in Agbogbloshie, Ghana.

The dispersal patterns of contaminants from e-waste burn sites are critical when these are located adjacent to agricultural land that provides a staple food source for residents of e-waste hubs (Davis et al., 2019), as is often the case. Agricultural soil is a vital and nonrenewable resource that performs food production and several critical environmental, social, and economic functions (Maliszewska-Kordybach et al., 2009). Heavy metals and POPs released from e-waste burning can penetrate soils where vegetables and crops are grown, transported by contaminated irrigation water and through direct deposition by air (Fu et al., 2008; Huang et al., 2011). These pollutants not only reduce crop yields, but can result in adverse human health consequences through the consumption of contaminated food (Wu et al., 2019). Plants can absorb these chemicals by their roots, transport them upwards to their shoots, and accumulate them inside their tissues. Direct foliar uptake of contaminants from the atmosphere can also occur during plant growth (Bi et al., 2009), with significant variations in uptake among different plant species (Luo et al., 2005). Studies on the accumulation of heavy metals from rice and organic pollutants from vegetables grown near e-waste incineration sites have shown high levels of lead, cadmium, polybrominated biphenyls (PBBs), PBDEs, and PCBs in local food samples (Fu et al., 2008; Zhao et al., 2009). In animal and fish products, there is the added threat of bioaccumulation or bioconcentration in the food chain (Fu et al., 2008; Luo et al., 2011). For example, Song et al. (2011) investigated levels of PCDD/Fs in fish, duck, hen eggs, chicken, rice, and vegetables collected from Taizhou and found the levels of PCDD/Fs in foods of animal origin were at least thirty times higher than those in vegetables and cereals, which exceeded the maximum limit standards set by the European Commission.

Oral ingestion of contaminated food has been shown to be a particularly important pathway for transferring these contaminants from the environment to human bodies. Yu et al. (2019) analyzed the multiple pathways, including ingestion, inhalation, and dermal contact of heavy metal exposure for residents of an e-waste hub in Qingyuan,

China. The results found that the consumption of rice and vegetables with high concentrations of heavy metals was the primary exposure pathway to both non-cancer and cancer risks. In a related study, Tang et al. (2015) analyzed mercury exposure in Taizhou, China. They found a strong positive correlation ($p < 0.01$) between hair inorganic mercury and time staying in the industrial area ($r = 0.81$) and between methylmercury and fish consumption frequency ($r = 0.91$). These findings imply that workers were mainly exposed to mercury vapor through prolonged inhalation of contaminated air and dust, while nonworker residents were primarily exposed to methylmercury through high-frequency fish consumption. The estimated daily intake of mercury showed that dietary intake was the primary mercury exposure source and mercury intake from rice and fish was significantly higher than any other foods. Ohajinwa et al. (2019) conducted a similar study assessing the public health risks of e-waste recycling in Ghana but sampled e-waste workers exclusively and analyzed both heavy metals and PBDEs (which served as a proxy for the cocktail of POPs emitted in e-waste hubs). Here, dermal contact was the main route of exposure for e-waste workers, while the cumulative health risks via all routes of exposure (inhalation, ingestion, and dermal contact) exceeded the acceptable limits of both non-cancer and cancer risks. Taken together, these studies have led to calls for increased attention to occupational safety programs and management regulations for e-waste workers (Ohajinwa et al., 2019), attention to food safety, avoidance of vegetable cultivation in regions with e-waste burning (Wang et al., 2012; Yu et al., 2019), and remediation of e-waste contaminated sites – particularly former incineration sites (Awasthi et al., 2016; He et al., 2019; Luo et al., 2011; Soetrisno & Delgado-Saborit, 2020; Yu et al., 2019; Zhao et al., 2015).

It must be emphasized that the results of e-waste contamination studies are locale-specific, with a systematic bias toward certain climatic and biogeographic contexts. While the overall picture shares many commonalities with the results reported in the international literature, in the West Line villages, there is far less use of acids, which tend to mobilize heavy metals, the lithology is almost entirely sedimentary, and rainfall is limited in amount and season, shaping the chemical composition, dispersal, and uptake patterns in unique ways. As we will see, these factors become of overwhelming importance in tailoring remediation to each context.

7.2 A Convenient Omission

Nearly every media report, activist Facebook page, grant proposal, or policy report on e-waste in the Israeli and Palestinian context opens with a dramatic yet perfunctory listing and often visualization of toxic sites (ARIJ, 2012; GLSHD, 2013; Shammai, 2017). Sometimes, these reports are inflected by and harnessed to a political message, whether decrying Israel's "toxic colonialism" and "e-waste dumping" in the West Bank, or the reckless danger of allowing Palestinian control of areas over "Israel's" aquifers. Seldom do these contain much about how the legacies of these environmental crimes are to be remediated. And, as evidenced in Extended Producer Responsibility (EPR)-based management systems discussed in Chapter 3, government policies are exclusively forward-looking. This enables governments and environmental planners to propose management plans that purport positive economic gains by collecting and extracting value from e-waste items, with little regard for the cost of past contamination embedded in the environment. Indeed, Israel's EPR policy begins practically at the moment of the law's implementation and conceptually when e-waste was "discovered" as an overlooked resource, whose value could be captured. Here and elsewhere, the notion of "responsibility" in EPR policy does not grapple with the substantial toxic legacy of e-waste mismanagement.

This disjunction has social, disciplinary, temporal, and practical roots. The shapers of policy are almost never those living in contaminated hubs, so their concerns, time horizons, and investments in change are different. E-waste is (and is seen to be) generated in urban developed contexts, with hubs in peripheral undeveloped ones. The skills and epistemic horizons of scientists who map and investigate toxic hazards are different from those of the policy or economic experts who deal with e-waste. Moreover, the cleanup of existing sites must, necessarily, be deferred until burning and improper processing have ceased – so that in a purely temporal sense, cleanup is a "secondary" issue. Also, as daunting as the challenge is of reforming the manufacture, disposal/collection, and recycling of e-waste, the scale and complexity of remediating existing sites is, perhaps, equally large and complex, and without the dangling opportunity of value recapture. But for those living in the hubs, we would argue that the stopping and remediating tasks are deeply linked and equally important. In fact, this linkage is one of the main motivations for our call for a hub-driven approach to e-waste policy, discussed in length in Chapter 5.

A disregard of the toxic legacy is evident in the emblematic case of China, which has played a massive role in global e-waste flows and management, and forceful recent efforts to reform these, and has a vast legacy of sacrificed e-waste processing areas and a correspondingly extensive literature and expertise on e-waste contaminants and contamination sites. Over the past decade, the Chinese government has drafted a series of policies and regulations to control pollution resulting from e-waste dismantling by reducing the number of informal e-waste recycling workshops (Lu et al., 2015; Ni et al., 2013). As a result of these policies, local governments constructed formal e-waste disposal and dismantling centers and strengthened enforcement to eradicate informal e-waste recycling activity, resulting in many abandoned contaminated sites (Wu et al., 2015). In Taizhou, the local authority made efforts to reduce contamination in agricultural soils. Since 2012, hundreds of illegal e-waste dismantling workshops have been banned, while some were merged with a larger industrial park (He et al., 2019). The city also strengthened industrial emission standards, promoted green technology, and expanded the priority regions for comprehensive protection (Yang et al., 2020). Yet, remediating contaminated soil has not been an integral component of China's government actions, so that abandoned e-waste disposal sites continue to pose ongoing harm to the surrounding environment (Lopez et al., 2011; Zhang et al., 2014). For example, Yang et al. (2020) analyzed the outcomes of China's pollution control policy interventions on the concentrations, sources, and health risks of heavy metals in soil in Taizhou, which was the center of environmental regulation and promotion of control measures intended to lower pollution levels. Their results showed that lead levels had declined in soil samples, yet cadmium, copper, nickel, and zinc concentrations were not affected. Wu et al. (2018) also analyzed abandoned e-waste burn sites in China after the government cracked down on informal e-waste recycling in the Guangdong province and found heavy metals had dispersed from the site into nearby water bodies. These studies concluded that due to their persistence and bioaccumulation in soils, remediation of contaminated soils is required to improve soil quality and protect human health.

7.3 The Challenges of Remediation

Remediating contaminated sites presents massive and complex challenges. They have the severity of industrial contamination, but also the dispersed distribution of non-point sources. In sheer chemical terms, the challenge is one of the most severe imaginable, with multiple dispersed sources releasing an unknown and variable composition of heavy metals and organic contaminants, through multiple and shifting dispersal and uptake pathways (Chen et al., 2019; Chigbo et al., 2013). Whereas organic contaminants bioaccumulate but can break down over time, heavy metals are toxic in their elemental state and do not break down, though they can be rendered biologically less active when bound to substrates. If burn sites were only contaminated with heavy metals, remediation would likely follow one of two strategies. The first option is to remove heavy metals from the soil through physical, chemical, or biological methods so that the content of residual heavy metals in the soil is close to background values (Liu et al., 2012; Yao et al., 2012). This method is effective, although the cost is quite high. The second option is to stabilize heavy metals in the soil through chemical passivation, that is, adding passivators into the soil to lessen the mobility and bioavailability of heavy metals through a series of reaction mechanisms including adsorption, precipitation, complexation, and ion exchange (Huang et al., 2016). This approach is much easier and at a lower cost than the first option, enabling large-scale soil remediation (Ma et al., 2020). However, stabilizing heavy metals may not deliver the desired results in co-contaminated soil within an acceptable duration, and its performance is highly dependent on the surrounding environment (Gutierrez-Gines et al., 2014; Ijaz et al., 2016). Thus, in situ soil stabilization technologies, while less expensive, are likely inadequate to remediate e-waste-contaminated soil, leading to recommendations for removal and ex situ treatment of contamination, particularly for incineration hot spots that contain a diverse mixture of high contaminant concentrations (Zhang et al., 2012).

An added factor is that some of the more complex sites of industrial contamination occur in peri-urban settings, while e-waste hubs are often in more rural ones where the real-estate value of brownfield recovery is not a motivation for remediation. Policy also must grapple with divergent local standards and technical capacities for hazardous waste treatment and the cost and distance to an approved long-term repository for the most severe fraction of the ash and soil removed in remediation.

Finally, not only are the remediation constraints and needs different in each regional locale, they can also differ from site to site at a micro level. At a regional level, soil metal toxicity and bioavailability depend not only on the total metal concentration in the soil but also on other physiochemical parameters such as pH, organic matter content, and soil texture. For example, in contrast to e-waste hubs in China and India, the West Line villages do not conduct harmful acid washing since the population has limited access to chemicals required to carry out these processes. Here, circuit boards are exported rather than leached locally, which shapes the kinds of chemicals released as well as the acidity levels that tend to mobilize metals. Also, the climate limits dispersal. Thus, when Friedlander et al. (2019) analyzed soil samples collected from an e-waste incineration site in East Jerusalem for the presence of heavy metals (i.e., iron, manganese, copper, lead, and zinc), in contrast to previously discussed studies in tropical climates, they found that only copper and lead were at concentrations above the 2018 US EPA screening levels for residential and industrial soil. Moreover, transect-driven sampling indicated that heavy metals did not disperse more than ten meters beyond the visible edge of burn site ash, which could be attributed to aridity (which is even greater than the West Line sites). The authors suggested that heavy metals released by uncontrolled burning in semiarid environments may be limited to exposure from very fine particulate matter or exposure to smoke and direct physical contact with ash or dust from within the burn site, so that remediation could be achieved by merely collecting, removing, and safely disposing of burn site ash.

A later careful reanalysis of the East Jerusalem site and an initial analysis of an adjacent one showed how location-specific contamination patterns can be. At the micro level, a sampling of this site based on local micro-topographic features showed considerably longer-range water-borne transport. Analysis of another site, only a kilometer away, where battery casings were ground, found lead levels an order of magnitude higher, as well as massive levels of arsenic, but without the telltale visual indication of black ash. A series of detailed site analyses in the West Line showed how highly local micro-hydrological and substrate characteristics gave rise to surprising differences in transport from outwardly similar sites, with elevated levels of water-borne contaminants over a kilometer in two sites of prolonged burning, and levels below detection only tens of meters away in another.

7.4 The Design and Derailment of the West Line Remediation Plan

While e-waste contamination in the West Line villages arises not only from burning but from burial or dismantling processes, burn sites are a primary and visible source of sustained contamination. For this reason, our community-based research in the West Line villages underscored the urgency and importance of removing burn site ash from the villages to restore the landscape and reduce exposure risks. As part of our efforts to facilitate and produce a comprehensive proposal to reform the West Line e-waste hub, we developed and tested plans for a feasible remediation strategy.

We estimated that the total cost to remediate all the most serious e-waste burn sites in the West Line villages would likely range from $3–15 million dollars, depending on the chosen remediation strategy. Our plan was restricted to a tentative $1 million budget from the Swedish International Development Agency (Sida), so we focused on treating the cores of the most severe sites, with more extensive and detailed work to be done if additional funding from the same or another donor and/or government agency became available. Based on our mapping of the location and activity of burn sites, initial characterization of their contaminants, and analysis of overland dispersal by air and surface flow, we believed that while some contamination had spread beyond the immediate sites of burning, the most benefit would be gained by removing surface ash from these sites and storing them locally or at a more distant hazardous waste facility. In developing a burn site remediation plan, we followed the same hub-centered principles outlined in Chapter 5, favoring local governance, democratic decentralization, and support of those segments of the e-waste industry whose livelihood would be hurt by interventions.

In line with the e-waste remediation literature, consultations and site visits with Israeli, Palestinian, and international hazardous waste management experts offered two possibilities for contaminant removal and storage. The first option, preferred by experts coming from other contexts, was for in situ confinement and storage, that is, removing burn site ash to be buried in plastic-lined sinks containment (either at

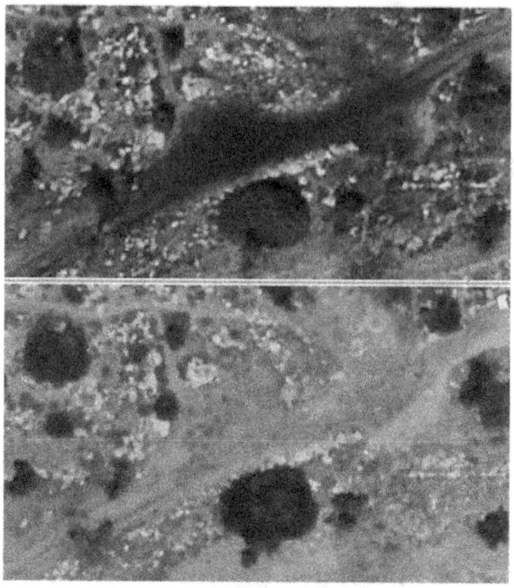

Figure 7.3 Two satellite images showing the pilot remediation of a burn site in Beit Awwa. Image (a) shows the burn site before remediation (2015) and image (b) shows the same burn site after remediation (2016)

each burn site or in a series of waste confinement storage locations serving several sites nearby). Notably, this is the most common remediation strategy for Superfund sites in the United States. The second option is ex situ treatment of burn sites, which would remove and transport contaminated soil and ash for bioremediation at Ramat Hovav (Israel), the only hazardous waste treatment facility in the region, serving Israel and, when possible, Palestine. Each of these options has its benefits and drawbacks. Put simply, in situ treatment is much less expensive but poses an ongoing ecological threat.

To determine the costs and logistics of in situ treatment, our team conducted a pilot remediation of two burn sites, under the guidance of two international experts (see Figure 7.3). Based on this, we determined that seventy to 100 sites could be remediated within our $1 million budget. However, while in situ treatment would remove hundreds of tonnes of contaminants from the surface, these would remain buried throughout the West Line villages – requiring diligent tracking and avoidance of these buried hazards in perpetuity. Ex situ treatment is considerably more expensive and complicated. We were unable to conduct a pilot remediation for ex situ treatment but were able to estimate costs and determined we would have funds to remediate ten to twenty burn sites. Ex situ treatment also would require Israeli and

Palestinian approvals at the national level to transport hazardous waste over a contested international border.

Regardless of the remediation method, only a portion of all burn sites could be cleaned, given our financial limitations. This raised another important consideration: Which burn sites should be remediated first? In consultation with our hazardous waste management experts and community partners in the West Line villages, we identified the following considerations to be important when selecting and prioritizing sites for remediation: risk to human health and proximity to local residences; the size of the site and severity of contamination; geological context and vulnerability to aquifer contamination, agricultural land use, topography, and runoff characteristics; the cost of cleanup and distance to hazardous waste landfill/storage site; local buy-in and commitment from landowners to prevent recontamination of cleaned sites; symbolism/visibility of prominent locations and equitable attention to various clans/families.

Two of these factors deserve some explanation. In geological and hydrological terms, a mapping of the cumulative activity of burn sites in relation to the recharge area of the Mountain Aquifer and local geological features showed that the greatest long-term damage would be due to the subset of sites to the east of the villages, as well as those carrying contaminants to karstic interfaces with subsurface flows. In social and political terms, prioritization of cleanup would need to consider the structure of clans or families in the West Line villages to balance benefits and avoid the appearance of favoritism or even corruption that might be triggered by a cost–benefit design of cleanup priorities and scheduling based sheerly on public health and environmental considerations. As discussed in Chapter 4, in this area even municipal leaders, employment ratios, and council memberships were carefully selected to ensure equal representation from families, and we tried to take such factors into account in all three arms of the project (livelihood transition, local enforcement, and remediation).

A final decision for a proposed remediation plan was determined through discussions with the West Line community representatives as part of our Delphi method, among other stakeholders in the West Line villages and the Palestinian Environmental Quality Authority (EQA). The mediation of health and safety criteria, local preferences, and national-level bodies – in this case, the Palestinian EQA – was productive and at times challenging. For example, the EQA insisted that not only should contaminated ash and soil be removed, but also that clean arable topsoil should be placed on top of the vacated areas, so that farming could resume right away – a reasonable request incorporated into the next iteration of our proposal.

Regarding the question of in situ versus ex situ scenarios for disposal of the removed toxics, things were more complex. West Line residents and burn site landowners preferred in situ remediation since it would remove more contamination from the villages and offer a more significant risk exposure reduction. This was despite the symbolic-political allure for some residents to send the hazardous remnants of Israeli e-waste back to Israel, thereby demonstrably purifying the Palestinian land. The Palestinian EQA firmly favored the latter, both as a more technically robust method as well as the appeal of orchestrating the transfer of hazardous waste back to Israel after what they regarded as decades of being helpless in the face of Israeli waste "dumping" in the West Bank (Aloni, 2017). While the "dumping" narrative does not map cleanly onto the e-waste sector, where Palestinians work hard to import e-waste into the West Bank, the symbolic justice of the ex situ transfer was an overwhelming factor for the EQA, despite reducing the number of sites that could be cleaned by almost an order of magnitude. Having witnessed the volatility (not to say chaos) of governance and land-use decisions in the Palestinian context, and with the hope that an initial more limited but thorough cleanup would open the way for the funding of its expansion, we could also see a certain logic in the more expensive and therefore limited extraction scenario for the initial Sida project. (Somewhat ironically, as described below, it was problems raised by this transfer of toxics to Israel that brought the project to a standstill.)

In addition to the site selection aspect, our cleanup proposal specified some additional elements of the cleanup effort's overall institutional envelope. Once burn sites were selected, a call for bids would need to be tendered, and the winning companies contracted in a way that assured the scope and quality of remediation and worker safety. An unconventional element of this contract would be that insofar as possible, and without compromising effectiveness and accountability, the cleanup effort should favor the employment of local workers, especially those that might suffer from other elements of the project, such as e-waste burners. Finally, a repository of the location and records of all cleared sites and stored contaminated material would also be created and turned over to the local municipalities, the

Palestinian EQA, and the Ministry of Local Governance to guide later remediation efforts and land-use decisions.

While the ex situ plan involved significantly fewer sites and removed the need for arranging local long-term interment of the contaminants, it required transporting hazardous waste over a contested international border. As discussed in Chapter 3, the international movement of hazardous waste is regulated by the Basel Convention – an international agreement to which both Israel and Palestine are signatories. To transfer hazardous waste, the Basel Convention requires state-to-state agreements through a prior informed consent process. In line with their aspirations to become an independent state, the Palestinian EQA has demanded adherence to the Basel Convention protocol for hazardous waste trade, which requires formal agreements with their counterpart in Israel – The Israeli Ministry of Environmental Protection (MoEP). However, Israel does not regard these protocols as applying here. As we were told by officials in the Israeli MoEP, "the Basel Convention is between states, and, with all due respect, Palestine is an Authority, not a State." With the assistance of a former Secretary of the Basel Convention, an expert in international environmental law, we made some progress in mediating a workable compromise on this issue. Thus, Israeli officials agreed to a transfer arrangement for the waste that would not be between states but appear to be identical to the Basel Procedure in terms of the documents exchanged and the fact that the exchange would be between the respective point people for the Basel Convention in both Israel and Palestine. Initially, it seemed that this would be acceptable to Palestinian environmental officials.

A further hopeful move to break the deadlock was made when we persuaded the Israeli Civil Administration to request a grant from the Israeli Ministry of Environment Cleanliness Fund (an omnibus budget line fed by hypothecation of the deposits on most recyclable containers) that would cover all the costs on the Israeli side. Thus, Sida would cover the cleanup and transfer of waste to the border, and Israel would cover the transport from the border to Ramat Hovav and the costly processing and internment of the waste at the facility. This arrangement both boosted the leverage of the limited budget and carried symbolic valence since, from a Palestinian perspective, Israel could be seen as "taking responsibility" for its waste. On the other hand, the Civil Administration's interest in undertaking the remediation of several burn sites themselves was rejected on the threshold. While couched as a desire to protect Palestinian health and environment, the move was, with good reason, seen as a poorly veiled extension of Israeli jurisdiction into the West Bank – an apparently benevolent form of soft creeping annexation that would underscore the inability of Palestinian institutions to protect and govern their own areas while legitimizing Israel's land-use interventions in Palestinian areas. The Palestinian EQA crisply rejected this proposal.

Despite the hard-won progress toward a plan that could be funded and begin, further tangles emerged. From Israel's point of view, all civil matters related to the occupied Palestinian territories are managed by the Civil Administration, which has divisions for each of the spheres of life that would fall under the jurisdiction of the respective relevant ministry. Thus, there is an environmental division within the Civil Administration with both military and civilian employees that claims to have ultimate authority over all environmental matters within the territories. On the other hand, while the Palestinian Authority (PA) interacts ongoingly with the Civil Administration, the EQA would not agree to the environmental division of the Civil Administration as their counterpart for the transfer of waste. They insisted that their counterpart be the Israeli Ministry of Environment – that is, this was to be a transaction between states. This deadlock was made clear during a discussion on September 7, 2017, between members of our team and the Palestinian EQA, when Yasser Abu Shanab, Deputy Director General of Environmental Protection, EQA, declared:

> I mean all this is because they are already a member [of the Basel Convention]. Now the problem does not lie with us, it lies with the Israeli party. They do not want the coordination to be through the Israeli MoEP, they want the coordination to be through the Civil Administration, and we refuse that. . . . in the Basel Convention we are dealing state to state . . . we need to regulate all the issues from the official ministry in Israel, and through our ministry. This is the official way to do this process for the transportation of e-waste.

While a *modus vivendi* for cooperation between Israeli and Palestinian officials has been worked out in some spheres – notably regarding security coordination, this was not the case for the transfer of waste, where trust had been eroded to the point that informal agreements with the Israeli MoEP or the Civil Administration were not possible. As we moved between the two sides, there was a Rashomon-like conflict between the perceptions and accounts of each. The Israeli environmental officials declared their ongoing history of taking toxics to Ramat Hovav and working to stop the transfer of waste to the West Bank, and their Palestinian counterparts recounted a series of betrayals and disappointments with respect to these same issues. Discussion with the then Chairperson of the EQA, Adalah Atteereh, made clear that she did not trust Israel to keep the toxic ash once it was shipped to Ramat Hovav, implying that the ash would ultimately find its way back into the Palestinian territories. From her perspective, these transfers could not take place until Palestine had control over its borders and thus able to block the dumping of waste.

One final factor was the locus of enforcement. Based on the effectiveness of the local policing efforts that had grown organically from the alert and response unit we had operated during the pilot stage of the project, our proposal had emphasized local ownership and location of the policing arm of the grant. The EQA, on the other hand, saw itself as the sole authority for all environmental enforcement and insisted that all resources and authority reside with their central Ramallah office. Our team and the local municipalities sensed that this would considerably dilute the effectiveness of this arm of the project, but reluctantly agreed.

At some point, however, the accumulation of sticking points grew too large. The national/municipal tensions, the Basel arrangement's implied reinforcement of Palestine's non-state status, and the inability to transfer waste to Israel despite an insistence on ex situ treatment of the waste proved too much and the project faltered, and Sida was forced to suspend its offer for funding, which had progressed so very far.

7.5 Concluding Remarks

In the Israel–Palestine context and globally, toxic sites are powerful symbols and advocacy tools for changing e-waste policies, yet are seldom practically addressed in proposed e-waste policies. We have no complaint with forward-looking policies, but when many burn sites now carry the long-standing residues of over a decade of absent or failing policies, they must be attended to as well. While policies that ban hazardous waste exports and encourage long-lasting electronics without toxic components and EPR-based e-waste management may reduce improper waste disposal and burning going forward, they fail to address the "slow violence" that will leach from existing sites for generations to come (Nixon, 2011). Those living in and around hubs are most sensitized to this half of the equation and must insist that futures not be separated from legacies in e-waste policies.

8 Border Frictions

Formalizing the Palestinian E-Waste Industry?

Though Ali lived in Beit Awwa, like many Palestinian men his age, he had spent most of his adult life working in Israel – twenty-five years in the construction industry. Like many of these men, he lost his job in 2002 during the Second Intifada, when the work permits of many thousands of Palestinians were revoked. Ali and the many newly unemployed men in the West Line villages now faced a suddenly bleak economic landscape: They could no longer rely on their place in Israel's "industrial reserve army" (Portugali, 2013), created by years of restrictions and de-development under Israeli occupation that constrained a weakened Palestinian economy. With limited construction and high unemployment, he struggled to find a construction job in which he would earn, at best, one third of his income in the Israeli context.

Instead, he took inspiration from his relatives, who were earning a respectable income in a newly booming entrepreneurial sector of collecting and dismantling e-waste from Israel. By the time we met him a decade later, he was a familiar character among the hundreds of e-waste businesses in the West Line. Unlike the lucky few who retained permission to enter Israel themselves, Ali had to partner with an Arab-Israeli resident permitted to drive in Israel, who could purchase cheap e-waste and scrap metal on his behalf and transport it back to the West Line villages. Initially, these were small loads of one to two tonnes, which he dismantled beside his home and then sold the extracted metals to local traders. The low overhead and quick turnaround on profits facilitated rapid growth, and he then invested his personal savings in a dismantling scrapyard in Idhna. Over time, he built up his facility by investing in machinery, hiring seven workers to dismantle e-waste and purchasing land to accommodate larger quantities. For a decade, business boomed: He imported upward of sixty tonnes of e-waste each month, with sales ranging from ₪200,000–250,000, at times more than what he used to earn as a contractor in Israel. While he was aware of the health and environmental hazards of his business, he enjoyed being free of the stress and daily grind of traveling to Israel and appreciated the rewards of his spontaneity and entrepreneurial spirit.

The prosperity of Ali's dismantling business began to encounter new frictions in 2014 as Israel grew increasingly aware of and began to clamp down on the transfer of waste, including e-waste, to the West Bank. At the end of 2013, the Israeli Ministry of Environmental Protection (MoEP) and its West Bank arm within the Civil Administration established the David Unit, which would operate at the borders to the West Bank to prevent waste trade, targeting construction waste, trees for charcoal, and e-waste. In its first two years, its staff of a dozen had apprehended and confiscated more than 300 trucks and requested $8 million for an additional thirteen staff positions with the Parks and Nature Authority, who partnered on this initiative.

These developments were happening in a parallel universe that the e-waste sector and other actors within the West Line villages were not aware of until they were surprised by their practical consequences. After a decade of hassle-free importing of e-waste and scrap metal into the West Bank, Ali's trucks began to be stopped by the David Unit and then began to be confiscated. Each time, the trucks were impounded for several weeks and released after payment of a fine of ₪7,000, which together with the storage fee and lost work time would set Ali back $10,000, a loss that would have easily taken a smaller business under. The ad hoc and shifting nature of the regulation added to the difficulty. There was little guidance as to what items were allowed or not, when confiscation was applied, and for how long and under what terms, and how trucks were released – often involving some shady forced sale of the load to an Israeli scrapyard at below-market prices. This kind of organized chaos is typical of Israel's interface with the West Bank, and that of the Civil Administration in particular, which, on the one hand, can operate remote from Israel's core governance institutions and under an aura of unquestionable security considerations, and, on the other, aloof from the Palestinians, to whom they seem to have no accountability. At a certain point, the challenges of informality exceeded the benefits, and Ali, along with many other scrapyards, became interested in formal operation, if such options were available.

Palestinian businesses that do manage to survive such border challenges do so through resilience and inventiveness. Ali and other successful scrap importers have learned to navigate new hazards, finding the less risky times and locations for crossing back from Israel. The rare occasions when his trucks are apprehended at the border can therefore be managed as a painful but survivable cost of doing business. The David Unit confiscated sixty waste trucks (not only e-waste) in 2020, which is a blow, but not a crippling one. A slower imperceptible process began at roughly the same time, as Israel's e-waste legislation (described in Chapter 3) came into force in 2014, bringing the shadowy realm of the e-waste economy into greater visibility while also making it more firmly illegal. The siloing of knowledge ran both ways, as the Israeli MoEP shaped the legislation with little knowledge of the Palestinian informal sector. In rendering Israeli e-waste into a national resource, legally owned by two Producer Recycling Organizations (PROs) funded to collect and recycle Israeli e-waste, the legislation formally rendered all other e-waste actors and transactions illegal. For e-waste recyclers like Ali, the development and implementation of Israel's Extended Producer Responsibility (EPR) law went unnoticed for years, and he only became aware of it when we discussed with him its implications for his business.

Informality is as much the rule as the exception in the West Bank where nearly half of all businesses operate informally (Kawasmi & White, 2010). For more than a decade, every evening about eighty trucks had returned with e-waste from Israel, passing through a central border crossing without resistance or questioning from border officials, who were more interested in people and loads entering Israel than leaving it. And, even with legislation, the categories and processes related to waste are still murky (which items are considered e-waste? How does one distinguish between used items that are intended for resale or refurbishment from truly end-of-life items?). Such questions are debated globally, as the heterogeneous and rapidly evolving composition of e-waste items complicates simple binary categorizations (Kahhat & Williams, 2009; Lepawsky, 2018; StEP, 2014; Widmer et al., 2005). How much more so in the unique case of transfer to the Occupied territories. Here there are multiple crossings, under different institutional structures (the army or private security companies outsourced by the Ministry of Defense's Crossing Authority) and a geopolitical situation in which Israeli-plated cars can cross over with little supervision and move freely around the whole of the West Bank. This is a result of the legal jumble of the Protocol on Economic Relations (Paris Protocol) that was signed in 1994 for an interim period of five years and is still patchily in force, and the intermittent and strained coordination between Israeli and Palestinian officials. Not surprisingly, the Israeli MoEP and Palestinian Environmental Quality Authority (EQA) struggle to regulate travel and diverge in their interpretations of almost everything.

This was the context in which we worked to upgrade and gradually formalize the West Line e-waste industry, which represents the third pillar of a hub-driven approach to reform the West Line e-waste hub, alongside strategies to reduce e-waste burning (Chapter 6) and remediate toxic sites (Chapter 7). The broad goal of the reform pillar was to allow entrepreneurs such as Ali a viable alternative to either operating in a furtive and environmentally destructive manner or abandoning the business and livelihood he cultivated over years. He and others like him were willing to register their businesses, pay taxes, and invest in the infrastructures needed for clean operation, if this would assure a stable, legal way of operating, especially to import their inputs from Israel. For Ali, formal operation promised to reduce the risk, open access to loans and insurance, and eliminate the Israeli middlemen parasitic on the informal status and restricted access and mobility of Palestinians. Thus, hub-driven reform entails facilitating a supportive environment for informal recycling businesses to transition to profitable, yet environmentally sound operations.

The remainder of this chapter details our efforts to develop, test, and promote a plan for this kind of gradual formalization. We describe the emerging regulatory context on both the Israeli and Palestinian sides, which is a key shaping factor, and our efforts to gain the support of key stakeholders and decision-makers from Israeli and Palestinian environmental authorities through extensive consultations and shuttle diplomacy. We describe two stances or possibilities that loomed large as potential pitfalls: a simplistic attempt to stamp out rather than reform the industry, and the more subtle but equally worrying scenario of a facile formalization that would reinforce rather than relieve the power polarities between the formal and the informal, Israel and the Palestinians. We describe the unexpected allies and adversaries discovered through these efforts, and the significant legal, political, and logistic barriers encountered. Against this overall background, we describe our piloting of sector transition through one company – a major cable-grinding facility, Safa Recycling, one of the only formal recycling businesses in the West Bank, which we partnered with to demonstrate the potential for a transformed e-waste industry.

9.1 Coalition Building and Sensitive Framings

In the spring of 2017, the David Unit confiscated Ali's truck, sending ripples through the industry. Our discussions with him and other concerned scrapyard owners centered on how they could operate formally to avoid such aggravations at the border. Ali asked us directly: What would it take for me to operate a formal e-waste recycling business? He was willing to install a concrete foundation and drainage system in his scrapyard to prevent leakages from the dismantling area, to grind his cables instead of burning them, and to adhere to other environmental requirements. But what were these?

As we worked to answer this question, we learned that the EQA does not have codified environmental standards for e-waste recycling; in fact, there was no formal e-waste recycling system in the West Bank at all. Similarly, Israel's e-waste law did not list specific requirements for e-waste recycling facilities, which are guided by the general requirements for treatment facilities. These are detailed in Section 38(1) of the law and include: (1) scales for weighing waste; (2) impermeable surfaces; (3) covering to prevent the waste equipment and batteries from getting wet; (4) suitable storage means for dismantled spare parts; (5) collection receptacles suitable for different components of waste equipment and batteries, including batteries or accumulators, transformers containing polychlorinated biphenyls (PCBs) or polychlorinated triphenyls (PCTs), hazardous waste and radioactive waste; (6) a separate system for the drainage of effluents from the impermeable surfaces; (7) a facility for the treatment of effluents; (8) treatment facilities must keep and make accessible a monthly inventory of e-waste entering and leaving the facility.

It would be well within Ali's reach to upgrade his facility to meet these standards, and we met many more scrapyard owners who shared the same desire to transition to formal and legal operation. This meshed with the development plans articulated by our community representatives from the West Line villages, described in Chapter 4, who agreed that their local economy is well situated to build on the existing e-waste industry's strengths.

Two major obstacles stood in the way of these hopes for formalization. First, Israeli law restricts e-waste processing to Israeli citizens and residents. Second, the law contains no direction for how hazardous scrapyard residue and waste materials are to be treated. Representatives from MoEP determine a facility's final approval, and the requirements may vary between plants. The MoEP has made no real attempt to inspect or enforce recycling standards during operation (Shammai, 2017), and some approved Israeli facilities operate in a manner not that different from their Palestinian counterparts.

Against this background, we began to explore modalities through which the West Line e-waste hub might continue its *de facto* operation as a massive collection and processing system, but in a legal and profitable way. Over the course of 2015 to 2017, we iterated a proposal in this vein, bouncing the ideas and written statements off relevant authorities and influential stakeholders, seeking both to inform and gain insights and reframe the issue of this "pirate" sector in ways that built support for an alternative conception that would coincide with many of their own interests. On the Israeli side, this involved the MoEP, Civil Administration, and PROs to determine a procedure and set of standards for Palestinians to collect and transfer e-waste from Israel to the West Bank, and formally integrate Palestinian recyclers with Israel's e-waste EPR system. On the Palestinian side, this involved the EQA as the body that must permit e-waste trade, set and enforce e-waste recycling standards, and potentially support a request for economic incentives to support formalizing recyclers, such as short- to medium-term value-added tax (VAT) waivers. The municipal governments in the West Line villages were also important stakeholders since they are closer to the day-to-day operations of scrapyards and more likely to ensure they adhere to recycling standards. Finally, environmental nongovernmental organizations (NGOs) in Israel and Palestine already engaged with the e-waste issue were also important stakeholders to ensure that their advocacy message to Israeli and Palestinian decision-makers aligned with ours.

This effort was demanding, not only in terms of shifting stances, misinformation, and overcoming institutional fragmentation and competition within each side, but, even more so, in moving between these alienated and mistrustful sides. While the Israeli and Palestinian e-waste systems are inextricably linked and both in a period of great dynamism and pressure for intensive reform, the processes are disjoint and at times even antagonistic to one another. The entwined geographies and economies required – and could benefit from – shared management, but the issue also held the potential to fall hostage to the respective combative political positionings and programs. The dominant Israeli view was to forget its years of complicity in benefiting from cross-border trade that conveniently collected and disposed of its e-waste and frame the sector as reckless and uncontrollable, or even as a form of "chemical warfare" that needed to be stamped out, a rationale for extending Israeli control over the Palestinian areas that had threatened Israeli aquifers by failing to govern themselves. The dominant official Palestinian view, on the other hand, was to ignore the great efforts and initiatives of the informal recycling sector, and its outsized contribution to the Palestinian economy (Davis & Garb, 2018a), and frame the phenomenon as Israel's toxic dumping, due to Israel's undermining of the Palestinian Authority's (PA) ability to control its borders and the extensive Area C lands in which most burning took place.

Our advocacy tried to navigate carefully between these antagonized and antagonizing framings and actors, shuttling back and forth with information and perspectives and searching for "trading zones" (Galison, 1997), concrete ways in which both sides might adopt a measure as beneficial to them, even if grounded in divergent framings. We distilled the core goals and interventions for all three pillars of the project into a one-page project matrix, in Hebrew, Arabic, and English, respectively, which we used within the communities and in our stakeholder meetings with donors and international actors, as well as the Palestinian and Israeli sides to ensure that our conversations were all anchored in a unified narrative. The initial meetings with stakeholders, in which most were appraised of the scope and scale of the issue, were typically structured around the following three elements:

(1) *You have a big problem.* We opened with a broad overview of the West Line e-waste hub, emphasizing in quantitative terms both its economic importance and its dire environmental and health consequences. Over 60,000 tonnes of Israeli e-waste had been collected and processed in the West Bank every year for over a decade, which generated more than 1,000 jobs and over $28 million in

gross value added (GVA) annually, but also residues of burning roughly ten tonnes of e-waste components daily (Davis & Garb, 2018a, 2018b). This overview included, in passing, the careful procedures through which we came to these figures. Our goal here was to establish a factual basis and ourselves as being credible to provide this, as well as jolt people into realizing the scale and severity of the phenomenon – that this was not a marginal issue that could be swept under the rug or deferred. In almost all cases, there was considerable surprise, even shock, when our presentation was made for the first time, as there was little existing information about the scale and severity of the issue, especially in robust quantitative terms. This is common with e-waste hubs throughout the Global South, with little solid empirical data on the flows and economies of even the notorious ones. Thus, only a few media and NGO reports have attempted to quantify e-waste hubs in China (South China Morning Post, 2002; Xinhua Net, 2005), Nigeria (Manhart et al., 2011; Puckett et al., 2005), and Ghana (Amoyaw-Osea et al., 2011; Prakash et al., 2010).[1]

(2) *Stamping it out by force probably will not work.* With this factual foundation established, we moved on to the second element of our presentations, which was to preempt what would have been the immediate response in most cases: eliminate e-waste ills through policing, bans, and enforcement. We suggested that, while this approach might seem appealing, it was likely to only drive the informal sector further into the shadows, weakening its capacity for clean operation and pushing it to relocate to less regulated areas and times. We based this assertion on examples of local scrapyard owners' ability to creatively evade or accommodate enforcement, as well as examples of ineffective government crackdowns in other e-waste hubs, primarily in China. In these cases, efforts to eliminate e-waste recycling sites have largely resulted in a scattering of polluting enterprises to operate in more covert conditions and less regulated locations – leading to a predictable "race to the bottom" dynamic (Chi et al., 2011; Goldstein, 2017; Zuev, 2018). Trade bans that restrict e-waste importation have also

[1] The flows and economies of informal e-waste hubs are difficult to quantify, and we find a range of methodological problems in existing studies that attempt to do so. For a systematic critique of previous studies, see Davis, 2017; Davis & Garb, 2018a.

encountered limited success. Thus, since the ratification of the Basel Convention in 1992, e-waste trade between developed and developing countries is supposed to be controlled and regulated, and many national e-waste trade policies have taken a step further by adhering to the Ban Amendment to the Basel Convention, which creates rigid binary regulations that strictly prohibit the movement of hazardous waste (including several e-waste components) from developed to developing countries (Lepawsky, 2015). Yet, as evidenced in e-waste hubs in Nigeria, China, India, and other sites, e-waste importation restrictions have done little to reduce inputs (SBC, 2011; Wang et al., 2020). Instead, they have merely altered the way that e-waste is shipped, for instance, instead of shipping a container labeled "electronic scrap," containers will be shipped as "reusable electronics" suitable for refurbishment and reuse, or e-waste will be shipped through other creative practices, such as packing used cars full of old computers.

With porous borders in many places and daily flows to the Israeli settlements embedded in the West Bank, the Israeli-Palestinian context offers its own unique advantages for creative evasions of regulation. Because Palestinians are not permitted to drive in Israel, they hire Arab-Israeli residents to drive their collected e-waste from Israel into the West Bank, and there is no system in place for checkpoint security to determine the final location of their trucks. For example, e-waste loads could be *en route* to a settlement, as several settlements host formal scrapyards, or to scrapyards in Palestinian villages. Moreover, several settlers have a hand in the reverse arc of this industry: They profit by facilitating the sale of extracted metals and circuit boards back into Israel. This blunts the possibilities of policing and motivates settlers to turn a blind eye to the traffic and burning around them and to advocate for the continued transfer of e-waste. On the other side, the PA might be very motivated to enforce a ban on cross-border e-waste trade, but it is in no position to do much about it, since it can neither supervise border traffic nor exert authority in large areas of the West Bank.

Just as the geopolitical situation complicates policing e-waste at border passages, it also limits control of dismantling workshops and burning sites in the West Line villages. The former are in Areas A/B, under Palestinian civil control, and the latter in Area C, under Israeli civil and military control. In the first case, Israel only enters under exceptional circumstances necessitating military action, and the PA has little reach in remote peripheries, such as the West Line, especially when the industry is so much part of the social and spatial fabric. In the second case, unless it involves a security threat, Israel intervenes only begrudgingly and belatedly in Area C, while the procedures of coordination needed for any action by the PA are so cumbersome as to render this option irrelevant. And in both cases, the authorities are overwhelmed with what they consider more pressing issues. When public or media pressure did prod the authorities to raise the policing pressure, burners began to drift to the south and east of the West Line villages and operate at night. Additionally, scrapyards have a light infrastructural footprint and can bounce back after demolition or relocate, as has occurred with the move of some West Line businesses to East Jerusalem villages. This kind of response was documented in China in 2002 after the Basel Action Network publicized their exposé documenting the severe environmental consequences of a massive e-waste hub in Guiyu (Puckett, 2002). Efforts from the Chinese government to ban such practices achieved little success as the industry's agile nature enabled recyclers to evade detection by working in remote and less policed areas, inside homes, and/or at night (Chi et al., 2011).

If policing at the destination and transfer points is difficult, what about reduction at the source? In principle, Israel's EPR law made e-waste the legal property of two Israeli PROs, to be tightly registered and managed through a system of contracts with businesses, importers, and local authorities on the one hand and certified dismantling facilities on the other, so that the more e-waste collected and treated in Israel by this formal system, the less will be available to recyclers in the West Bank. However, as in many other places, the long-standing agile, pervasive, and industrious collection network of Palestinian waste collectors in Israel has impressive advantages over a new formal system, especially one that was shaped in ignorance of the existing informal networks and incentive structures. Formal e-waste collection and recycling systems throughout the Global South have consistently been undermined by pervasive and preexisting informal sectors (Akenji et al., 2011; Ciocoiu et al., 2011; Ciocoiu, 2012; Kojima et al., 2009; Manomaivibool & Vassanadumrongdee, 2011; Scheinberg et al., 2010). In the Israel-Palestinian case, with the "developing country" a half-hour drive away, the pressures to leak material away from formal systems are even greater. These difficulties led us to our third claim.

(3) *Joint engagement and formalization are the best option for a deeply unified system.* Given these scenarios, we argued that the Israeli-Palestinian e-waste system is likely to continue operating as a unified one, certainly in the short-to-medium term and most likely beyond that. This led to the third and final element of our advocacy message, which suggested that the most viable path to improvements in e-waste management is a coordinated strategy to integrate the informal and formal, Palestinian and Israeli actors in a system with incentives and pathways to move to sustainable and increasingly formal operation.

We began with our own starting point, namely the motivation and willingness of the West Line businesses and community to move in such a direction, and then sketched the economic, environmental, and governance benefits likely to appeal to the wider audience of formal commercial actors, decision-makers, and officials on both sides. For example, both governments stand to benefit from new taxation revenues from a massive export industry now occurring through cash purchases and the facilitation of Israeli middlemen. If it led to reduced burning, cooperative efforts would benefit both the local environment and the aquifer underlying the contaminated area. Palestinian recyclers would benefit from a horizon of regular and dependable e-waste inputs underpinning jobs for thousands of workers, a consideration to which the Civil Administration was quite responsive. Thoughtful reform would also rationalize the faltering EPR e-waste management system on the Israeli side that is hampered by an informal industry that already exceeds collection quotas the formal system is struggling to reach, and this without subsidy. Reform would allow the system to tackle segments of the waste stream that cannot be economically recycled in Israel, but which could be recycled in the Palestinian context – as long as this did not compromise health and environmental standards. Thus, Israeli PROs and formal recyclers stand to benefit, or at least remain unaffected, from such an arrangement. We emphasized that these measures were to be gradual and that the goal was to set in motion a virtuous circle of standards, performance, and benefits. Paradoxically, at least when viewed from the pragmatic level of the hub, the crazily fused Israeli-Palestinian context actually offered singular advantages for a binational e-waste management system. However, what made sense and seemed, despite challenges, to move fast at the local West Line level, rapidly got messier and harder to advance once others with a less direct stake exercised their considerable influence.

9.2 Stakeholder Friction: The Interface of Hub Realities and National Legislation

Curiously, as a pair of outsiders, we could help shape and then advocate for a development path that was broadly consensual among many stakeholders within the West Line. This convergence, acceptance of our role, and local enthusiasm for the proposed package gave us confidence in our shuttle diplomacy to link other stakeholders and their interests to the proposal, fleshing out concrete details as we went. We should have suspected that as we moved outward from the hub, and those whose income and everyday health would rise or fall with the fate of the e-waste sector, other motivations and rewards would begin to swamp those from whom the project emerged, and things would get messier. But we did not expect the extent to which this would occur, nor that some of the stickiest problems would come from the environmental agencies and NGOs we thought would be easiest to persuade.

Thus, our first circle of outreach was the closest one to the West Line municipal governments of Idhna, Deir Samet, Beit Awwa, and Al Kum.[2] They were supportive and even proactive, rapidly shifting from being "stakeholders" to being partners in the efforts. This was not surprising, since our Delphi community engagement method (Chapter 4) had included representatives from the municipal governments. The notion of environmental upgrading of the existing recycling

[2] From 2013–2016 the municipal governments of Beit Awwa, Deir Samet, and Al Kum were amalgamated into a unified municipal government called Yassaria by the Ministry of Local Government, which was part of a process to make municipal governments more efficient and increase their access to national funding sources. However, due to the deep-rooted tribal differences between the villages, formal protests took place in Al Kum and Beit Awwa as the mayor of Yassaria was from Deir Samet and was believed to favor policies that benefited family members in his village. These tensions eventually led to the separation of the municipality of Yassaria back into the three smaller municipalities of Beit Awwa, Deir Samet, and Al Kum.

hub was instantly recognized as a way to reconcile their dilemma of constant complaints about burning on the one hand and the need to preserve livelihoods on the other. They realized full well that a blanket banning of e-waste by Israel (or, they soon realized, Palestinian authorities) would leave them with the worst of both worlds – elimination or relocation of the backbone of their economy and nothing done to remove e-waste processing's toxic legacy.

But the municipalities are beholden to the PA via the Ministry of Local Government, which has the authority to overhaul municipal decisions and can even appoint municipal leaders. Municipalities could do little, therefore, to shape or go against broader policies, so their efforts were restricted to those local ones that did not antagonize the national politics. For example, the Idhna municipality subsidizes the tipping fee at the Tarqumiya waste disposal facility to reduce the open burning of non-valuable waste.

At the national level, the Palestinian EQA has been hesitant to support a regulated cross-border e-waste trade since it requires legalizing Israeli e-waste importation. Officially, the EQA based this policy stance on their understanding of the Basel Convention and Palestine's Environmental Law, established in 1999 before e-waste was a prevalent issue. Our discussions with the EQA revealed that their initial interpretation of the Basel Convention was a strict ban on the transport of e-waste from Annex IIV countries (e.g., Israel) to Non-Annex IIV countries (e.g., Palestine). While this blanket ban might have made sense in some contexts, in the Palestinian context, it was merely declarative, meaning that making the import of used electronics illegal, whether for use or recycling or not, did not stop the flows at all; it only eliminated the possibility of any formal player from engaging with it.

We facilitated several rounds of discussion on the topic, including expert consultation with the former Secretariat of the Basel Convention, Katharina Kummer Peiry, through which we clarified several provisions in the Convention, such as the use of Prior Informed Consent, which would enable the EQA to abide by the Basel Convention and permit e-waste imports from Israel, if desired. However, it then became clear that this potential way forward was also blocked. The Basel Convention serves as a supplement to national laws addressing hazardous waste trade, and Palestine's Environmental Law explicitly prohibits the importation of hazardous waste, including under a new proposal, e-waste regardless of condition or whether it will be repaired or reused. Should this new proposal go into effect, any bilateral agreements to permit e-waste trade would require amendments to this annex of Palestine's Environmental Law or the law itself – a long process, even if there was the political will. Coupled with the overall EQA narrative of Palestine as Israel's "digital dumpsite," this annex more or less closed the avenues for formalization on the Palestinian side.

Though Israeli environmental personnel were not immediately aware of this pending legislative tightening, the Palestinian refusal to import e-waste limited their options too. While Israel does not recognize the PA with respect to international law, Israel is a signatory of the Basel Convention and therefore has agreed to prohibit the export of hazardous materials to non–Organization for Economic Co-operation and Development (OECD) countries. It could not afford to be stigmatized as a party to "toxic dumping" in Palestinian areas. And besides, a complete ban on export to Palestine also appealed to some actors in the Israeli recycling industry, who decried the leakage of waste, which the rhetoric around the EPR legislation had heralded as a treasure and the source of new green jobs, rather than a burden. Others in the industry knew, of course, that much of the flow was far from a treasure, with low and in some cases negative value which Israeli facilities could not capture and were more ambivalent about closing off leakage from which they benefited. And while those in charge of regulating cross-border flows knew firsthand of the difficulties of shutting off such permeable borders, it was easier to prohibit all waste transfers than to regulate based on the content and destination of each load.

These pressures from both sides to cut off the legal transfer of e-waste for recycling made any model of sustainable formal sector tenuous. And the EQA got caught up in the contradictions between the realities of enforcement and their nationalistic aspirations and declarations. They understood the environmental and livelihood consequences of the e-waste industry and that they could not enforce the ban they were declaring. The EQA also understood that while Israeli Authorities could, as they pointed out, be very effective in stopping items that posed a security threat, they had little interest or resources for enforcing e-waste flows. With public pressure and donor funding now aligned toward the "sector reform" model, they settled on a funny compromise: They would support the establishment of formal recycling facilities, but these would be restricted to "indigenous" Palestinian waste, not illegally imported inputs from Israel.

9.3 Surviving (and Thriving in) Ambiguity

Thus, our advocacy efforts were inserted into and prompted this situation of emerging ambiguity regarding the status of e-waste. In parallel to the clarification of the legal status of e-waste just described, and even as this structural regulatory blockage became clear, we engaged with a range of key stakeholders to explore the options for the win–win formalization trajectory we had initially envisioned. Here, we detail some of this engagement in these early phases of the Swedish International Development Agency (Sida) project. Our description tapers off at the point at which the EQA stance and regulatory framework eventually became too intractable for the Sida project's operational phase to begin, but that does not imply that the tensions and ambiguities we describe have been laid to rest. While momentum for the project was broken, many of the same questions continue to roil discussions of and stymie movement on this issue today, even as we write these words. As we will see, while the growing ambiguity and stasis at the macro level killed the effort for a grand formal integrative structural solution for the sector, they left the informal sector more or less untouched. The situation even provided a fertile ground for the emergence of some actors along an NGO–EQA axis, who found ways to bypass the political deadlock posed by explicit Israeli-Palestinian integration, on the one hand, and the local community on the other, so that they could pick up and obtain funding for some of the elements we had proposed (policing, formal operation) without the encumberment of pushing for the bigger vision.

On the Israeli side, we began a dialog with the two PROs, E-community and MAI, established under Israel's e-waste EPR law. Both conveyed an initial impulse to cut Palestinian recyclers off from Israeli e-waste. They were struggling to meet their mandated quota under the EPR law to collect 50 percent of the e-waste stream by 2021. The Palestinian sector, whose extensive and efficient collection network had historically collected something approaching this volume without any subsidy, made it difficult. They perceived the Palestinian sector as unwanted competition and initially advocated for the Israeli

MoEP to enforce a ban on cross-border e-waste trade. MAI, whose management's political and business models were somewhat more aggressive, went as far as making an appeal on this issue to the Israeli High Court of Justice in 2016, which resulted in the MoEP clarifying barriers designed to exclude the informal e-waste sector. Thus, informally sourced e-waste will not be recognized toward quotas, even if an accredited recycler can produce proof of its origins, and residents may be held criminally responsible if they dispose of their e-waste to an unauthorized collector (Shammai, 2017).

The hostile stance from the other PRO, E-community, loosened somewhat over several discussions in which we presented the same facts and perspective described above. Not only did the practical difficulties of cutting off e-waste flows to the West Bank become clearer, but they were intrigued by the possibility of benefitting from the established Palestinian collection network to rapidly achieve their collection quotas. They expressed real willingness to explore and possibly forward the options for formal integration of the Palestinian e-waste sector. And in May 2017, E-community joined us to meet representatives of the scrap sector in the West Line villages, who were interested in formalization that could regularize and stabilize their situation. During this meeting, E-community representatives even forwarded options for a special border zone in which transfer and dismantling activities could occur under the law. One of the stumbling blocks, however, was the clause in Israel's e-waste EPR law that made it illegal for noncitizens/residents to be employed in the collection and dismantling of e-waste. This clause, adopted before the dominance of Palestinian actors was understood, was a measure to preclude the sector from becoming dependent on immigrant labor, as construction and agriculture had become.

The Civil Administration's initial policy stance toward the West Line e-waste hub was also one of elimination of waste pirates, which threatened groundwater and air quality. In our discussions though, as they became aware of this system's size and stability, as a recycling sector rather than marginal pirates, attitudes shifted, and there was greater willingness to engage. Several things contributed to this. For one, our study indicated that the industry directly employed at least 1,000 men, and for the Civil Administration, unemployment meant unrest, especially in the Hebron Governorate, which they regarded as especially unruly. Another factor was the limited effectiveness of their own intensive and expensive efforts to reduce e-waste burning through forceful policing alone. The presence of David Unit staff and a million dollars of video camera coverage at the borders, the occasional dispatch of military jeeps to large burn events, and demolition of several scrapyards located in Area C, made some dent but did not really stop the continuous complaints of burning from Israelis near the border or in settlements. The final element was the effectiveness of our brief pilot project, described in Chapter 6. Burns dropped to 20 percent of their former levels, reported by army observers as well as Israeli residents, who now reported a summer without smoke, in which they could sit on their porches without the recurring nuisance of being forced to retreat inside.

This combination of their own costly measures with little impact and our external measures to reduce burning that had a marked effect got their attention in a new way, and they became willing to discuss the other elements of the project – formal integration and site cleanup. Until then, we had been in a vicious circle, in which they had said "stop burning and we will discuss support for clean recycling businesses," and we had said "if you boost clean recycling, burning will diminish." The burn reduction pilot broke this deadlock, and we began more practical discussions of permitting cables to a pilot cable-grinding facility, and modalities for an eventual cleanup of burn sites, and transport of hazardous soils and ash to a hazardous waste site, which we discuss in the next section.

Ironically, interactions with another set of stakeholders – environmental NGOs – proved thornier. As discussed in Chapter 3, Palestinian NGOs based in major cities outside the West Line villages (e.g., Hebron, Ramallah) fiercely opposed notions of regulating the existing e-waste trade. Having thin familiarity with the details of the e-waste industry and narrow exposure to the range of viewpoints within the villages, it was easy for these NGOs to align with global "dumping" narratives and the segment of locals more unequivocally opposed to the industry in any form, and to play up the dumping as a face of Israel's occupation. This framing immediately cast our proposed system integration as perpetuating Israel's oppression and promoting "normalization" (this despite the NGOs' own past participation in cooperative Israeli-Palestinian grants).

Obviously, in a broad framing, the conditions for the cross-border e-waste industry were forged in the context of enduring and repressive occupation and sustained by asymmetries of power. But there is no reason to think that those in the West Line hub were any less aware of these wrongs than those working in NGO or EQA offices in Hebron or Ramallah. What is different is the extent to which their lives are directly affected by the outcomes. The measure of success and institutional survival for an NGO is often a funded proposal, and for the EQA the ability to demonstrate their vigilance in protecting Palestinian autonomy and land or a newly opened factory. On the ground, the proposal or the opening ceremony is only the means to an end, which is change that will shape whether one's children suffer an asthma attack from smoke or will have money for schoolbooks or university studies.

Initially, the impulse of these Palestinian NGOs was to shut down the pirate industry, or bring some higher-tech elements of it to urban centers. But with the wide floating of our model for broader and in situ hub transformation and the success of the cable-grinding pilot project, proposals were submitted for similar facilities in the West Line itself. This, of course, faced the dilemma that the hub industries were almost entirely dependent on the now contraband imports of materials from Israel, no matter their destination. To avoid this, and in alignment with the EQA's declaration on this matter, these Palestinian NGOs said that their efforts were directed at cultivating green facilities that would treat domestic Palestinian waste only (ARIJ, 2012; GLSHD, 2013). Though local businesses know full well that over 90 percent of their materials are derived from Israel, whether directly or at one remove, they were willing to play along with this declaration if it enabled access to donor resources or other benefits.

9.4 From Bilateral Agreements to Pragmatic Solutions

In parallel to these developments, we proceeded to shore up and test the various elements of the overall project: the local policing discussed in Chapter 6 and the burn site cleanup discussed in Chapter 7. We were also pushing forward on a demonstration of the feasibility and effectiveness of a clean recycling facility, in the form of a cable-grinding company, based in Idhna. All the bilateral and legal issues just discussed still loomed. However, there was an existing (and EQA-certified) Palestinian company, Safa Recycling, described in Chapter 6, already struggling to do the right thing and eager to regularize the direct imports of cables from Israel as a way of cross-

Figure 8.1 The owners of Safa Recycling proudly stand in front of their cable-grinding machinery.

subsidizing and stabilizing its treatment of cables that were extracted from dismantled items (also brought from Israel, but of a far lower grade). While the broader issue of the legality of e-waste was still being hashed out, cables were in a less contentious category so that we could move forward on a demonstration project that might leverage broader change.

Safa Recycling began their business in 2013 when two Palestinian entrepreneurs from Idhna invested in high-end cable-grinding machinery from China (worth over $200,000) as a business venture based on an environmentally sound method to extract copper from cables (see Figure 8.1). One of the owners had spent more than a decade abroad and established the business upon his return, so that unlike many others, this was a large up-front investment, rather than incremental expansion. Perhaps because of this context, unlike the several hundred other businesses in the West Line, this one operated formally: It was registered, obtained all the necessary approvals, and paid taxes. While there was much fanfare at the ground-breaking ceremony of this first formal e-waste recycling facility, with attendance by the authorities and reports in the local media, the business faced a tough and lonely road as it began to forge a way of operating formally within an entirely informal ecosystem. Not only did they not get special support for their pioneering role, they also found themselves operating at a disadvantage to the informal businesses and even to other formal businesses in more familiar sectors. In the words of one of the owners, "it's the first one through the door that gets shot."

There was a string of regulatory barriers. At the time, and still, the Palestinian EQA does not have codified standards to certify e-waste recycling facilities but falls back on an ad hoc one-by-one assessment carried out by EQA employees. Even this was not easy to arrange, as the EQA is strapped for resources often; the only way to conduct the required inspections was for the owners of Safa Recycling to drive forty-five minutes to the city of Hebron to pick up EQA employees, who did not have access to vehicles, and escort them to their facility in Idhna. Despite these hardships, and after months of paperwork and inspections, Safa Recycling gained approvals from the municipality of Idhna, the EQA, Ministry of Health, Ministry of National Economy, and the Hebron Chamber of Commerce and Industry to operate their cable-grinding facility.

The next barrier, however, was that their business model was premised on grinding cables from two sources: local scrapyards from the West Line villages and direct importation from Israeli sources (see Figure 8.2). Based on preliminary discussions with local scrapyards, Safa Recycling anticipated a strong willingness to pay for cable-grinding services as an alternative to burning. But, when Safa Recycling opened for business, the cost of grinding cables, which ranged from \$180–550 per tonne based on the thickness and purity of the cables, was too much to attract the industry away from the familiar burning option.

This forced Safa Recycling to join the dozens of other businesses that sourced and imported cables directly from Israel, which were more homogenous and cleaner and so offered a higher profit margin. At the time, this import was a legal gray zone, but a few months later, in 2014, the Civil Administration doubled down efforts to ban the transfer of e-waste materials, and cables were the main target. Other businesses were undeterred, but as a formal business, Safa Recycling did not want to risk its status by getting caught smuggling cables over the border. Without cables to process, they reoriented their grinding

Figure 8.2 A scrapyard from the West Line villages delivers a load of cables sourced from Israel to Safa Recycling's cable-grinding facility. Here, Safa Recycling employees will clean, grind, and then separate the cables into copper and plastic pellets.

facility to separate copper and aluminum from air conditioner condensers that were dismantled by local scrapyards. This stopgap was not viable, as this material caused excessive damage to the grinding machinery while barely making enough income to cover operating expenses. This led to a crisis for Safa and a deterrent to all the other businesses that were watching to see how this first formal recycler fared before investing themselves in the upgrading and machinery they would need to begin to operate in an environmentally sustainable manner.

It was at this point that we began to partner with Safa Recycling to see whether it could obtain the Israeli approvals needed to import cables. We organized several meetings between Safa Recycling, the Civil Administration, and a sympathetic PRO-certified Israeli e-waste recycler that led to a working solution. Safa Recycling would first independently source and purchase Israeli cables. Upon collection, they would notify the Israeli e-waste recycler stating the quantity of cables collected, and for a nominal fee, the Israeli recycler would issue an invoice under an arrangement that "subcontracted" processing to Safa Recycling. The Civil Administration approved that these invoices would be presented to border security, allowing the safe passage of cables from Israel to the West Bank. Safa Recycling would then process the cables adhering to standards acceptable to both Israeli and Palestinian authorities. The 24/7 video cameras installed in their facility, with saved footage, was enough to reassure the Israeli Authorities that the imported cables were being ground, not burned. The valuable materials extracted by grinding (copper and aluminum) were then sold to formal Israeli or Palestinian recyclers based on a competitive price. Initially, the plastic outputs were sent to a landfill. However, our analysis showed that they contain a commercially valuable fraction of metals, so they began to be stored for a later point at which the machinery or partners were available to extract these metals and even recycle the plastic. All material outputs were accounted for with invoices matching the quantity of cables imported to Safa Recycling to reassure both the Civil Administration and the "subcontracting" Israeli e-waste recycler that the cables had not been burned. This arrangement commenced in 2015 and, barring a few minor misunderstandings with border guards, continued to operate successfully, continually being refined and expanded. Currently (the end of 2023), the facility grinds about ten tonnes daily, half of them sourced from Israel and half from the local scrapyards, thus redirecting many tonnes of cables from going up in smoke every day.

While this pilot project addressed just one segment of the e-waste stream and a single socially responsible Palestinian business, this was a high-leverage kind of waste and a configuration that offered a scalable model for other parts of the waste stream. For example, we began to elaborate a model under which Palestinian recyclers, in coordination with the relevant Palestinian authorities, could build upon their existing networks and procedures to operate as certified collectors or recyclers of Israeli e-waste authorized by one of the two Israeli PROs. Collected e-waste could then be transported to certified facilities in the West Bank, incentivizing businesses to upgrade their facilities to clean operation. E-waste would continue to be refurbished and resold on the secondhand market or dismantled to sell valuable metals to scrap metal traders in Israel or abroad, but with a playing field now shifted to favor formal and clean operation, with capital for investments needed to upgrade now accessible, and an incentive to dispose of non-valuable waste in certified landfills in the West Bank or Israel as a precondition for certification. By becoming certified to the same standards as Israeli collectors and recyclers, as well as those of counterpart Palestinian agencies, these collectors could begin to operate legally. They would incur the taxation and other costs of formal operation, but eliminate the risks of confiscated loads and the transaction costs of illegal operation (i.e., detours, paying middlemen). We also explored having these materials count toward the quota of the PROs, and whether some of the EPR subsidy revenues paid by the government to the PROs could be shared with the Palestinian e-waste partners.

The initial success of the Safa Recycling arrangement sparked considerable interest among others in the scrap industry. While they were dodging their smuggled loads across borders at unorthodox hours, Safa Recycling's certified loads were going through the front door – the official Tarqumiya crossing. But the long-term viability and scalability of this pilot effort were still vulnerable in two ways. The first was that the cable facility was able to defer some of the legal barriers to the cross-border transfer of e-waste because of the ambiguity of whether cables were classified as "e-waste." Discussions with Israeli-certified e-waste recyclers and PROs revealed that the cable's origin determines its classification as e-waste. For example, cables extracted from electronics or electrical equipment would be considered "e-waste," but industrial coils of cables, for example, would not. By assuming Safa Recycling is importing "non-e-waste" cables, Israel can bypass national and international laws prohibiting e-waste exportation to non-OECD countries. In reality, the cables Safa Recycling sourced from Israel and transported over the border might consist of a heterogeneous mix of both designations. Other waste streams would not be able to benefit from this ambiguity.

The second issue was that of sovereignty. Safa Recycling had obtained permission from the Israeli Authorities and operated during a pilot period during which these issues were being clarified on both sides. But eventually, Safa Recycling, and any other formal business, would have to interface with the stance of the Palestinian Authorities regarding imports that was clarified in our previously described discussions with the EQA. We had hoped the clean end-to-end loop of cable recycling demonstrated through Safa Recycling would shift the EQA stance on e-waste imports. Safa Recycling had obtained certification to recycle cables from the Palestinian EQA, but over time EQA policies solidified toward a sweeping ban on any kind of import so that any company seeking certification had to operate under the shared pretense that it would be processing only "Palestinian" materials. Since the PA has no control of border flows, whether formal or not, and Israeli enforcement was fairly leaky, other businesses could bring over materials illegally or draw on materials that others had imported in this way. But as a formal company that had explicitly established a legal arrangement on the Israeli side, Safa Recycling could not play this game. These issues put a clamp on any designs to scale this formalization model to other waste forms, and without the assurance of a path to legality, no business would sink their own funds into equipment to operate under adequate environmental standards. Thus, ironically, the EQA efforts designed to assert its sovereignty and advance environmental recycling and the rule of the law simply acted to perpetuate the operation of businesses in an informal and polluting status quo and to harm the single business that had invested in formalizing.

9.5 Concluding Remarks

While the story we have relayed is deeply enmeshed in the particularities of the Palestinian-Israeli context, it will resonate with experiences and tensions of the e-waste situation globally. These tensions include those between formality and informality, definitions of e-waste and regulations on the transfer of materials between OECD and non-OECD countries, between expanding local opportunity and national sovereignty, between the visions of hubs and those who would regulate them, and between declared policies and their actual consequences. The geopolitical and socioeconomic triggers that stimulate e-waste hubs' clustered location make targeted government investment and regulation feasible. Indeed, theoretical arguments and practical strategies to build the capacity of existing e-waste recycling facilities in the Global South are already being proposed and discussed in other settings (Gunsilius, 2010; Inverardi-Ferri, 2017; Oteng-Ababio et al., 2020; Rochat, 2008; Wang et al., 2012; Williams et al. 2013; Yu et al., 2010). We believe that the community-derived development path, the formal–informal partnership model, and the integrated three-pillar approach we forwarded all productively add to these discussions. In the Palestinian case, local politics have frozen this pathway for now, but these elements may take root in other less troubled places.

9 Open Questions, Opening Futures

Our deep dive into the emergence and functioning of the West Line e-waste hub has shown how this cross-border economy offers a perspective into the travails of the Palestinian-Israeli context, and how global environment and livelihood tensions and trends play out in this particular circumstance. In our closing chapter, we distill what we have learned from our past engagement in the West Line hub to project forward and consider how things might and should go in the future – both in the West Line and in the other e-waste hubs with which it shares many dynamics and predicaments.

An apt entry to this stocktaking might be the responses of the West Line business owners to questions we posed to them in a systematic survey we conducted in 2015 (Davis & Garb, 2018a). As part of our effort to quantify the volumes and values of e-waste flowing through the West Line hub, we established a full sampling frame of all e-waste businesses in the West Line, from which we drew a random sample of seventy of these stratified by business type (collectors, dismantlers, refurbishers, and traders), and in each interviewed the key figure using a semi-structured survey instrument. More open-ended questions at the end of these interviews asked e-waste business owners to look backward and forward at the trends and prospects of the scrap industry. Typically, these were senior people in the profession, able to reflect on the longer sweep of the industry and give an informed assessment of how the various forces at play might affect its future.

Most interviewees described a decline from the heady days of the mid-2000s. Then, metal prices were high, competitors few, environmental regulation (whether Israeli or Palestinian) almost entirely absent, and travel of Palestinians to Israel still unproblematic. The industry's informality meant it evaded government attention and was known only to those directly involved along with the suppliers and metal buyers in Israel and the Palestinians who bought their refurbished items. Speaking with them a decade later, however, they described how things had hardened. Many of the founding generation could no longer access Israel directly themselves because of tightening mobility restrictions. The commodities boom had cooled with copper prices down from their highs of $9,000 and even $10,000 per tonne to below $7,000. Increasing knowledge of the value and dangers of their industry had brought unwanted attention. On the Palestinian side, e-waste sources and strategies were no longer a guarded trade secret of few, and as scrap went from being merely rubbish to becoming a commodity, competitors were increasing. While still informal, the industry was no longer operating relatively free from government pressure, as the e-waste topic had been surfaced sufficiently for Palestinian NGOs and even central authorities to begin paying attention. On the Israeli side, NGOs and government agencies were taking action due to pressures from below (media-amplified complaints about environmental and health impacts from citizens and NGOs) and from above, with the upward harmonization of environmental regulation required as a precondition for Israel's accession to the Organization for Economic Co-operation and Development (OECD). In parallel, the Palestinian Environmental Quality Authority (EQA) had become increasingly aware of and engaged in the issue, undergirded by the legal infrastructure described in Chapter 7.

The macro-level regulatory shifts afoot on both sides remained mostly unknown for many years to those in the West Line e-waste industry – none of our interviewees were aware of them. Israel's draft Extended Producer Responsibility (EPR) law passed its final reading in the Israeli parliament in July of 2012 and fitfully progressed to approval at the start of 2014. This law stipulated collection quotas that would ramp up annually to a target of manufacturers and importers needing to recycle half the total weight sold by 2021. This EPR system combined with legislation banning the disposal of e-waste in landfills after 2020 provided the legal impetus and the rudiments of mechanisms and institutional structures that would redirect more and more of Israel's e-waste into formal recycling channels. In parallel, also under the impetus of its OECD accession, the Ministry of Environmental Protection (MoEP) had begun harmonizing its hazardous waste regulation to conform with OECD restrictions on the treatment and transboundary movement of hazardous substances.

Hazardous waste could not be exported from OECD to non-OECD countries at all, and non-hazardous wastes intended for recovery required more licensing than in the past. The question of transporting potentially hazardous materials and e-waste to the Palestinian areas, across the uniquely peculiar boundary constituted by the Green Line, did not figure in the periodic OECD assessments of Israel's progress leading up to accession, nor in Israel's response to these. But this upward harmonization held the potential to leverage regulation of the waste trade with Palestinian areas. Specifically, while the Basel Convention allows the export of materials for recycling from Israel to a non-OECD country if the recipient gives written consent, the Palestinian Authority (PA) can protest such transfers in the absence of such consent, as it has done for toxic wastes (Abdel-Qader & Roberts-Davis, 2018).

When we asked our interviewees how they saw the forces they mentioned (international metal prices, the occupation, the policing regime at the border, and their own agency in the face of these) playing out to affect their businesses in the future, we heard a mixture of confusion, fatalism, and resilience, often condensed into variations of a recurring phrase "Bas Allah Be'lam" (only God knows):

> You do not know what will happen in the future. You do not know what will happen in the global economy. In 2008 no one was working, demand was low, prices were low, it froze everything, but by 2011 we were all working in this industry again. Right now (2015), it is slowly dying. I am following the economic problem in Greece a lot because it directly affects me. I do not understand the whole situation. The worst seems to have already passed. Anything that happens in any country affects us. China affects us the most since they import 40 percent of the world's copper. [metal trader from Idhna]

> The Israeli Environmental Police is another main thing. They act irrationally. I expected them to take away my truck at the border and inspect it if they suspect I have cables. But what they did was to confiscate my truck – they did not inspect it! If I protest, "I do not have any cables – that is the thing that is prohibited by law" they respond, "I am the law." So, God knows what they will enforce or ignore in the future. [scrapyard owner from Idhna]

> If the border restrictions remain it will kill the industry. But this industry will never stop. My dad spent 28 years in this industry. It will fluctuate, contract and expand over time based on metal prices and accessibility. Only God knows the details. [scrapyard owner from Deir Samet]

Overall, they anticipated a continued decline over the coming five years but were confident in their own resilience to survive these worsening conditions. Indeed, as we engaged with the sector over the coming five years, we concurred with their prognosis of decline, resilience, and continued unknowability of what will unfold going forward. This continued unknowability is not surprising, not only because of the Israeli occupation and turbulent Palestinian/Israeli relations, but because e-waste is at the nexus of substantial issues felt in e-waste hubs more generally. The hub-driven model and policy package derived from the West Line villages at the center of the e-waste problem that would curb environmentally destructive practices, heal the toxic legacy, and gradually formalize and upgrade the sector remain inhibited by an array of external forces. While the outlook in the West Line and in general is challenging, if not bleak, the outcomes are not entirely determined. Much depends on the shifts in perspectives and policies to which this book aims to contribute. For this reason, the remainder of this chapter segues from surveying the forces of unknowns currently at play, to a more prescriptive orientation – drawing on our analysis of the West Line to advocate for what might be possible.

9.1 The West Line State of Play

9.1.1 The Geopolitical Arena

The multitude and depth of structural issues in the geopolitical arena that impinge on the West Line's future are staggering. The contours of Israeli/Palestinian economic and territorial integration or separation and the undoubtedly painful path to their resolution obviously lay down the overarching conditions for the industry's future. We can only sketch in the broadest terms the dynamics, with which more experienced commentators have grappled for over five decades, with no end in sight. In principle, within the framework of a two-state solution, which had been fitfully presumed for decades, Palestinian governance systems might set the tone and offer pushback against the unilateral dominance Israel has held for so long. However, decades of Israeli encroachment and spatial and infrastructural dominance, anchored in the settlement enterprise, steadily eroded the degrees of freedom for a two-state option. Indeed, it is increasingly argued that the spatial, demographic, and legal entrenchment of Israel's occupation has perhaps already foreclosed this, raising once unthinkable one-state or confederation geopolitical scenarios.

This was the situation in late 2022, the point at which we closed our narrative and sent off our manuscript, and from which we penned the paragraphs that follow. The Oslo Accords governing territorial governance and the Paris Economic Protocol governing economic relations were in tatters yet operational. The resultant ad hocism in daily life reflected and perpetuated the stasis and shaped the minutiae of the industry's practical operations: who approves solid waste transfer stations where, which taxes and import practices apply to whom, which trucks and loads pass and why, and so on. Our expectation was that in the coming years, we could expect the sector's fate to be altered not by a sweeping negotiated agreement but, if anything, by unilateral moves or events. At that point, international bodies, especially at the European level, were considering, at the PA's urging, making moves toward recognizing Palestinian statehood. The UN upgrading Palestine's status to a "non-member observer state" in 2012 was an indication of this trend. Aside from its apparent repercussions for the peace process itself, this designation would also shift the legal status of Palestine, with consequences for e-waste regulations. We also anticipated that on the Israeli side, a further rightward shift in Israel's next government would embolden the ruling coalition to move forward with aspirations several political parties have urged for some time now. These policies would unilaterally annex part or all of Area C, including the extension of Israeli environmental law to the annexed areas, thereby shifting the legal status of extensive areas in ways that would change the terrain of the e-waste industry. We also speculated that unexpected events such as resumption of hostilities or even the COVID-19 pandemic might limit flows to crippling levels.

The related questions of internal Palestinian governance were also key, and in flux. Even as we wrote in 2022, Palestinians were still waiting to vote for the presidential and Palestinian National Council elections for the first time after a fifteen-year suspension. Elections were initially scheduled for May 2021, but then indefinitely postponed in April 2021 due to uncertainties regarding whether Israel would permit Arab residents of Jerusalem to participate in the vote. We expected that if these elections were to occur, and the result were to be substantially different from the status quo, the shape of the PA might be quite different, both in its stance toward negotiations with Israel, as well as its internal authority and accountability. Were this shift to result in a considerably greater level of decentralization of authority and resources to local levels than has existed in the prevailing regime, this would open more space for the kind of hub-centered future we have suggested in the preceding chapters. Currently, with Israel's occupation as the key battle and terrain of struggle, the PA places great importance on centralization of its authority to create a unified oppositional front to Israel, which sometimes comes at the expense of the more differentiated framing required for a social and environmental upgrading of the West Line hub (see Chapter 9). It is understandable that in this stage of its emergence the PA argues that Israel's occupation is at the root of all problems. Yet, as long as this agenda overwhelms all others, the dumping paradigm with which it resonates so well is likely to dominate discussion of the industry, eclipsing local economic needs and foreclosing discussion of what kinds of scrap might be legally imported for profitable recycling.

But the PA's stridency against the industry (stigmatized as local mafia who collaborate with Israel's toxic dumping) is only declarative, with little ability to enforce the call to block all flows in and out. The outcome is the worst of both worlds – the blanket ban does little to lessen or upgrade the environmental performance of the informal trade, while blocking those who would operate formally or might upgrade their facilities. The PA's current efforts on recycling take place in this hollowed-out gap between declarative purity and practical impotence. For example, the EQA's recent approaches to donors for support of recycling efforts, or celebration of donor-funded projects for green recycling by Palestinian NGOs, adhere to the pretense that the inputs to recycling facilities are indigenous Palestinian waste, not originating in Israel. To the extent that formal e-waste businesses truthfully declare their purchases and sales to operate, they cannot build business models based on a charade that Israeli sources are not in the equation. Thus, they will be excluded from participation in governmentally approved recycling schemes, which will center on NGO operations, reliant on continued donor funding rather than viable commercial models.

Such a scenario would be likely to set in motion a cascade of consequences building a configuration opposite to the one we hoped to build, locking in a vicious rather than virtuous cycle. Clean recycling alternatives would survive only in the context of narrow performances for demonstration to funders, rather than offering real solutions for market forces and needs. Clean recycling remains a niche, not the broad basis of employment and profit that communities might point to in urging their family and neighbors to "do the right thing." This undermines the moral basis for local enforcement. And, in any case, if the EQA retains enforcement as its prerogative, which it directs at a distance to the geographically and socially distant peripheries where infractions occur, enforcement will lack a local operational efficacy as well. Absent effective community-based enforcement, burning will continue. Thus, there is little reason to clear toxic legacies, other than similar shows – limited symbolic operations for public relations goals, which, we have seen, are soon covered by new burns, or forgotten.

Thus, somewhat tragically, the PA's single-minded focus on augmenting and solidifying Palestinian autonomy locks in a vicious cycle that undermines this goal. In seeing the Palestinian scrap merchants as simply dupes of Israeli dumping and insisting that recycling operates only with local Palestinian inputs, they strengthen the standing of the Israeli recycling sector at the expense of Palestinian merchants who have dominated it for decades. By insisting that burning enforcement be done by the central authority and in the absence of the livelihood transition that would address the root causes of burning, they assure its ineffectiveness and increase the chance that Israeli actors will use continued burning and Palestinian inaction as a justification to "step in" in ways that further erode the extent of Palestinian territorial and governance autonomy.

Thus far was our speculative conclusion at the end point of our narrative, in late 2022. While our research and efforts on these topics continue, our account has to stop somewhere, and this end point is a natural one to seal our account, not only logistically, with submission of the book manuscript, but, also, as historical break point coinciding with the ascendance to power of a strikingly right-wing coalition in Israel. The subsequent acceleration and increased explicitness of decadal trends shifted the political landscape in significant ways, some of which had direct impacts on our topics of waste burning, cross-border mobility and trade, and the possibilities of forging the kind of potential cross-border arrangements we describe as possible. We can describe these only briefly, reminding our readers that the bulk of this chapter – and the book as a whole – remain, as we initially wrote them, just before the tumultuous year of 2023.

For example, over the course of the year, with a new government coalition now ranging from the right to the far right, the continual question of Israeli exposure to smoke from Palestinian burning shifted from being an environmental annoyance to being a readily weaponizable argument on behalf of a kind of "green annexation," which would extend Israeli environmental law into Area C and military operations into Area B. Many Knesset members and ministers, who had barely known of the issue, were readily mobilized on behalf of such interventions to solve this "environmental" crisis. Government agencies and the Civil Administration were under increasing pressure to demonstrate as forcefully as possible that they were doing something about it. The Gaza war that began in late 2023, and which is still underway as we update the manuscript's final submission, polarized the situation yet further. At the micro level, for example, immediately after the October 7, 2023, Hamas attack, a severe closure was imposed on Palestinian areas, with municipalities in the West Line and elsewhere now struggling to transport municipal waste to distant landfills, and a subsequent rise in ad hoc dump and burn sites. This led, in turn, to increasing settler complaints about what they now openly regarded as "chemical terror." In addition, with almost all work permits cancelled, unemployment rose steeply, leading to new levels of poverty and desperation. Thus, while it became more difficult to bring scrap from Israel, men were now willing to go to greater lengths to do so, adding to the burns. At a macro level, the structural contradictions of the occupation are now strained in a way that has created an even more tense and militarized interface between the West Bank and Israel, and raised the likelihood of more epochal shifts with respect to international intervention and the future of the PA, to the point where further prognosis is futile. Thus, the remainder of this chapter reflects on the forces underlying the situation at the late 2022 end point, and, especially as these might be generalizable to the contexts and dynamics of other hubs worldwide.

9.1.2 The NGO Arena

While the "hard" parameters of the operating landscape are likely to be forged in the geopolitical arena just described, the NGO arena has become important in shaping discourse and visibility about the waste sector. In addition to the NGOs we described in Chapter 3, others have more recently joined the fray. Their efforts, too, have not done much to brighten the prospects for the sector and, in some cases, have had the reverse effect.

For example, the Green Israel Forum, an Israeli environmental group closely allied with the settlement enterprise and militant right-wing NGOs, took on the issue with gusto. It mounted a clandestine surveillance campaign and used their tracking of trucks to the West Line area and images of waste burning as the basis of a sensationalist media campaign (e.g., a prime-time TV "exposé," dramatic YouTube videos), which foregrounded a simplistic nationalistic narrative about pirate scrap as a threat to Israel's national security. These campaigns laid the ground for legal proceedings, in which the NGO, with the apparent support of one of the Israeli Producer Recycling Organizations (PROs) (MAI), requested an injunction against the other PRO (E-community) and the Israeli MoEP for allowing this leakage of materials across the border. Fortunately for the Palestinian scrap industry, the legal proceedings did not produce the intended outcome as the court dismissed the claim against the government and E-community sued for defamation and won.

On the other end of the political spectrum, B'Tselem, a peace advocacy organization, wrote a comprehensive report on Israel's toxic dumping, entitled *Exploiting Palestinian Land for the Treatment of Israeli Waste* (B'Tselem, 2017). This report presented a series of case studies about sites in which a range of materials (sewage, medical waste, solvents, and an e-waste recycling facility on an Israeli settlement), but did not distinguish between those flows that are brought to the West Bank for extraction of value and those with negative value, which are sent there as a destination for cheap, illegal disposal. The EQA then took this report to confirm that the e-waste sector was a shady mafia operation, which overrode the national interest.

The more mainstream environmental groups in Israel frame the cross-border waste issue (including that of e-waste) "apolitically," that is, within conventional environmental terms of regulation and presumed shared interests in a clean environment. Their preoccupations are with Israel's internal environmental challenges so their engagement with cross-border phenomena is at some remove from the details and subtleties of these phenomena and from its livelihood and political dimensions, which are so central to their Palestinian counterparts. Their framings cast Palestinians as neighbors, who surely have common interests in protecting a shared environment, or render the Palestinian side into a neutral space onto which Israel's problems are "externalized," only to rebound back to Israel in the form of downstream cross-border problems.

At a local level, when smoke from Palestinian industrial or disposal burning comes over the border at many sites along the Green Line (and to Israeli settlements within the Palestinian areas themselves), frustration and motivations for concrete change are high. Residents in several locations have mounted pressure groups and campaigns to end this assault on their health. In the case of the West Line, the inhabitants of the half dozen neighborhoods a kilometer or two west of the Palestinian burn sites have complained and lobbied for intervention for years and are currently in the process of preparing a High Court of Justice injunction against the authorities. This will be aimed primarily at eliminating all cross-border flows (waste in one direction and extracted metals in the other), despite our efforts over the years to widen their considerations toward working jointly toward nonpolluting livelihoods, rather than the forceful elimination of polluting ones. Indeed, their interest is quite narrow and opportunistic – stopping the smoke – and they have expressed that they would be satisfied with Israeli enforcement if they could at least push the burning further to the east, away from their neighborhoods.

9.2 Global Discourses: Dumping and Destitution Versus Ingenuity and Resilience

Thus far, we have offered an update on the current state of play on where the context and dynamics of the book's case study stood in September 2022. The book has described our efforts to forge and advance a different way of moving forward, arguing that it had relevance not only for the West Line hub, but other toxic livelihoods too. The West Bank in the decade of 2011–2021 was hardly the easiest of contexts in which to trial a hub-based model of e-waste sector reform. The West Line model pushed not only against fixed conceptions of formal and informal, and of the sources of agency in and beneficiaries of environmental policy – challenging agendas in any context – but also, it did this when the forces against change aligned with some of the most entrenched power structures in the region. No doubt, this would and will be the case when a hub-based reform is attempted in other places as well: It is not accidental that various forms of political, ethnic, discursive, and market power are aligned. For this reason, while the local politics that so frustratingly froze our promising efforts are *sui generis*, they are still instructive. In the sections that follow, therefore, we keep in mind but at the same time abstract away the Palestinian-Israeli specifics of the West Line experience, to extract lessons and suggestions for building similar hub-based reforms in other contexts.

Resistance to the kind of bottom-up hub-driven change we advocate for will take local forms in each context. However, we can anticipate this will be imbricated with pushback against intrinsic and general challenges the approach poses to entrenched conceptions, power structures, and policy processes:

- The transition to cleaner livelihoods that we advocated valorizes the informal processes currently seen as marginal and pirate, or given mere lip service. We argue that their place at the policy table and within a "partnership" with the formal sector should reflect the fact that they have and continue to manage the vast majority of global e-waste.[1]
- Our model also argued (and then demonstrated) that enforcement should be local, and that its effectiveness was first and foremost a *consequence* of the availability of cleaner recycling practices, rather than a means of forcing the transition to these. It is only after livelihood options in a hub broaden to realistically include cleaner practices that local enforcement becomes morally and pragmatically supportable.
- We argued that visions of closed-loop futures for waste could not ignore the legacies of the electronic/electrical technologies that for decades presumed and repressed the existence of one-way extraction → use → disposal processes. We insisted that reforming the world's e-waste practices includes eliminating contamination sources and

[1] This figure is derived from The Global E-waste Monitor's 2019 estimate of 53.6 million tonnes of e-waste generated globally, of which 17.4 percent is documented to be "collected and properly recycled," 8 percent is estimated to be discarded into waste bins in high-income countries, and the remainder is largely unknown, although believed to be managed primarily by the informal sector. The total value of raw materials in global e-waste generated is estimated at $57 billion USD (Forti et al., 2020).

then attending to remediation of the sites to which their unmanaged products were conveniently directed for so long.

In other words, the hub-based model's challenge to conventional approaches for e-waste sector reform is likely to be, at the same time, a push against actors that have an interest in maintaining the current relations of power within broader force fields: between the formal and the informal; between local governance and state and global institutions; and between the temporal politics and investments of mobile institutions and communities and those that are place-bound. Keeping in mind these ways in which any hub-based approaches mobilize fundamental axes of politics, we can begin to parse the work ahead into spheres of engagement. Having barely emerged from our own case study, we can offer only a sketch of what we see to be the outlines of the discursive, institutional, and economic challenges at different scales.

A continual challenge is that of framing, which marks the hubs and its inhabitants as victims or pirates, roles that assign them to a future of elimination or, at best, a shallow "partnership." As we have seen, partnerships that emerge from the formal pole tend to dislodge informal participants from their sources and markets in ways that downgrade their position in the waste economy, receiving only the scraps of the scrap. Their only alternatives in such a partnership would be reactive: a slow, awkward migration away from an increasingly untenable niche or using their proven resilience and enterprise to dig into an even more invisible and deleterious mode of operating, making the designation of a "pirate industry" into a self-fulfilling one.

A concerted local and global effort is needed to counter the considerable rhetorical and institutional pull of this framing, to give substance to the reframing of hubs as remarkable agglomeration clusters, ripe and eager for social and environmental upgrading. The model of change must be with hubs as active – even driving – partners, rather than sites or recipients at the receiving end of imposed formalization processes. In fact, ideally hub-driven formalization should be (and be seen as) a side effect of upgrading rather than a coercive regulatory measure or precondition for upgrading. Formal institutions and procedures are a means for communities and businesses to choose to expand their options and resources.

Reframing is not a free-floating discursive endeavor, but part and parcel of legal, institutional, and economic changes on various fronts.

These include: (1) national-scale government and regulatory agencies; (2) the formal private sector recycling industry; (3) the international EPR paradigm and its regulatory correlates; (4) local and international NGOs, whether on the environmental or health side of the spectrum, or on the livelihood and development side, and their funders; (5) research and knowledge creation efforts; (6) businesses and actors (workers, owners) in the informal sector; and (7) the existing local and community institutions of governance and organization. Some of the changes and needs are implicit in our tale of the West Line efforts; others must be reinvented or invented. Here we can only give some starting questions, passing the torch on to our readers, who find themselves in the different locations and spheres in which efforts will be elaborated and tested.

A fascinating and strategic question is, "Where will the initiative for creative change come from?" The answer to this question will be different in each case. In the Israeli-Palestinian context, we expect the interface of Israeli PROs and those Palestinian private actors best situated and motivated to formalize will prove the most hopeful – and perhaps only – engine for change. They have the most pressing needs and skin in the game and cannot wait for the government regulator to change the playing field. Without some shift in the current situation, forward-looking Palestinian businesses have no path away from the dead end in which imports from Israel are increasingly encumbered by border logistics and, possibly, banned on both sides. The Israeli PROs are incentivized to somehow capture the "leakage" of materials to the informal sector, which undermines their business model and ability to meet quotas. Even were they to lay hands on this e-waste, they do not have a mechanism to introduce it into the accounting that is the basis for the subsidies they receive. Can the PROs find ways to engage with the informal sector's collection capacities (and not simply in a way designed to identify and displace them from their sources)? Can they identify segments and phases of refurbishment and dismantling best handled by each, and formulate workable mechanisms of accounting for and cross-subsidizing between the valuable and harmful elements of the waste stream? Not only is a hub-centered transition impossible without acquiescence of PROs, but they must also take on an active role in educating, lobbying, and reassuring national-level legislators that this kind of partnership is necessary and feasible.

National regulatory bodies are often underfunded and overworked – this is certainly the case in Israel. And, especially in smaller countries, regulatory bodies will be wary of diverging from the EPR templates on which they rely to experiment with the new logics and procedures required for engaging with this messy sector. Bureaucratic institutions often struggle to embrace and effectively execute an exploratory, incremental approach or turn a blind eye to the infringements of informal economic activity, necessary during a transitional period. For this reason, the global EPR paradigm, which sets the tone for national regulation in many places, is also due for an overhaul. It would be helpful to have insights and models emerging from local experimentation polled, pooled, and synthesized into tools and models that can expand the horizons of regulators and advocates internationally. We cannot be naïve though about the milieu and interests in which the current EPR model has evolved (some would say, was captured), which may be hesitant about extending the notions of "extended" and "responsibility" embedded in the EPR moniker in the ways we have described. Their orientation has been toward capital-intensive and formal reforms of e-waste management that neglect past e-waste contamination and run counter to the kind of needs of hubs in the Global South.

The communities, local authorities, and waste sector within hubs need to do much organizing if they are going to drive change. A broadly consensual vision for change must emerge from people differently located along the pollution/profit spectrum, and across the cleavages and constituencies that will prevail in most places. E-waste hubs must develop a unified voice, demands, and representative body with which to engage with the formal sector and local and national government. Good examples of such organizing exist in the waste-picking sector, but there is a curious absence of a trade sector organization representing the next tier up in the e-waste economy. In the Palestinian context, for example, the villages of the West Line hub produce more economic value and engage similar numbers of enterprises and workers than some of the celebrated sectors of the Palestinian economy (olive presses, souvenirs and handcrafts, or textiles) do nationally,[2] yet lack similar sectoral advocacy organizations.

[2] For example, in 2014 olive presses comprised 243 enterprises, employed 860 workers, and generated $9.15 million gross value added (GVA) to the national economy (PCBS, 2014c, 2014d). The retail sale of souvenirs, craftwork, and religious articles comprised 1,129 enterprises, employed 1,792 workers, and

Municipalities will often need to persuade and battle state-level powers to assert the economic importance and aspirations of legitimacy of the industries they house, and reclaim enforcement and revenue-gathering authority back to the local level. Were they to gain directly from the licensing fees and tax income from business formalization, for example, they would be more proactive about pushing in this direction. And, if democratically elected, they will be better able to balance the overall costs and benefits of formalization transitions.

Another arena, close to our own hearts, where shifts are due, is research and knowledge production. A hub-driven approach would require a shift in the locus, style, and substantive questions of inquiry into e-waste practices and policies. Currently, academic research is fragmented into three rather separate networks: management/policy, the chemistry of recycling and material valorization, and health/environmental impacts – apart from some recent ethnographic work, very little of this work emerges from the perspectives and needs of hubs. Even discussions of the informal sector, or analyses of the contamination taking place in hubs, do not seem to initiate from or be closely collaborative with these "objects" of inquiry. We have advocated for and, hopefully, offered a convincing example of a different and more participatory hub-centered style of research and shown that engagement with those closest to the pains and benefits of the industry yields different insights and directions for change.

We have mapped the multiple changing and unpredictable forces that will shape the future of the West Line and shown that many are shared with similar places elsewhere. And we have offered what we have learned from our efforts that might be helpful. Is the kind of "worst of both worlds" decline already in motion inevitable? Will this hub and others be driven deeper into informality over time, their options for upgrading blocked by short-sighted legislation, their sources of income slowly drying up, and contamination on the ground left untouched? Or are the hub-driven framing and set of measures we worked for with the community possible, if not in this messiest of contexts of Palestine–Israel in the 2020s, then perhaps elsewhere?

generated \$22.9 million GVA (PCBS, 2014e). The manufacture of textiles comprised 386 enterprises, 1,230 employed workers, and generated \$11 million GVA (PCBS, 2013). Each of these industries produces less GVA than the West Line e-waste industry, which comprises 381 enterprises, employs 1,098 workers, and generates over \$28 million GVA (Davis & Garb, 2018a).

Our experience offers evidence for both scenarios. The hub-based pilot worked. As we speak, ten tonnes of cables are directed daily from burning to grinding in Safa Recycling – a facility our hub-based approach helped set in motion nearly a decade ago that continues to operate now on a purely commercial basis. The local enforcement system we developed with local partners and piloted proved able to rapidly bring burning down to a fraction of its former levels. And by collaborating closely with community members from the West Line and sympathetic allies outside the villages, we demonstrated, costed out, and arranged the logistics and funding that would remove thousands of tonnes of contaminated ash and soil from the worst sites in which they accumulated over more than a decade of continual burning. And yet, that promising and proven community-driven pilot did not move further. The funding for its scale-up was left on the table by the politics invoked once the project moved beyond the consensual setting of the local hub communities in which they were forged and tested, and into the broader contexts of national and binational politics.

Is the hub-driven reimagining of e-waste futures we suggest for the West Line and similar hubs feasible or impossible? We have passed on our portrait and assessment of the situation, together with the resilient fatalistic assessment of our interviewees and companions in the scrapyards who are in the thick of things: Only God knows – Bas Allah Be'lam. We hope that these might provoke and inform our readers to join us in shaping the next stage of this drama.

References

60 Minutes. (November 2008). Following the trail of toxic e-waste. *CBS News*. Available at: www.cbsnews.com/news/following-the-trail-of-toxic-e-waste/

Abdel-Qader, S., & Roberts-Davis, T. L. (2018). Toxic occupation: Leveraging the Basel Convention in Palestine. *Journal of Palestine Studies*, 47(2), 28–43.

Abramson, A., Becker, N., Garb, Y., & Lazarovitch, N. (2011). Willingness to pay, borrow, and work for rural water service improvements in developing countries. *Water Resources Research*, 47(11), 1–12.

Acaroglu, L. (May 2013). Where do old cellphones go to die? *The New York Times*. Available at: www.nytimes.com/2013/05/05/opinion/sunday/where-do-old-cellphones-go-to-die.html

Ackah, M. (2019). Soil elemental concentrations, geoaccumulation index, non-carcinogenic and carcinogenic risks in functional areas of an informal e-waste recycling area in Accra, Ghana. *Chemosphere*, 235, 908–917.

Adam Teva Ve'Din (Israel Union for Environmental Defense). (2012). Recycling electronic waste: Knesset approves law. Available at: www.adamteva.org.il/?CategoryID=420&ArticleID=1479

Adeyi, A. A., & Oyeleke, P. (2017). Heavy metals and polycyclic aromatic hydrocarbons in soil from e-waste dumpsites in Lagos and Ibadan, Nigeria. *Journal of Health and Pollution*, 7(15), 71–84.

Adjei, A. (April 2014). Life in Sodom and Gomorrah: The world's largest digital dump. *The Guardian*. Available at: www.theguardian.com/global-development-professionals-network/2014/apr/29/agbogbloshie-accra-ghana-largest-ewaste-dump

Afenah, A. (2012). Engineering a millennium city in Accra, Ghana: The Old Fadama intractable issue. *Urban Forum*, 23(4), 527–540.

Agarwal, B. (2001). Participatory exclusions, community forestry, and gender: An analysis for South Asia and a conceptual framework. *World Development*, 29(10), 1623–1648.

Ahmad, M. S., & Talib, N. B. A. (2015). Empowering local communities: Decentralization, empowerment and community driven development. *Quality & Quantity*, 49(2), 827–838.

Akenji, L., Hotta, Y., Bengtsson, M., & Hayashi, S. (2011). EPR policies for electronics in developing Asia: An adapted phase-in approach. *Waste Management & Research*, 29(9), 919–930.

Akese, G. A., & Little, P. C. (2018). Electronic waste and the environmental justice challenge in Agbogbloshie. *Environmental Justice*, 11(2), 77–83.

Akortia, E., Olukunle, O. I., Daso, A. P., & Okonkwo, J. O. (2017). Soil concentrations of polybrominated diphenyl ethers and trace metals from an electronic waste dump site in the Greater Accra Region, Ghana: Implications for human exposure. *Ecotoxicology and Environmental Safety*, 137, 247–255.

Ali, M., Peerlings, J., & Zhang, X. (2014). Clustering as an organizational response to capital market inefficiency: Evidence from microenterprises in Ethiopia. *Small Business Economics*, 43(3), 697–709.

Aloni, L. (December 2017). *Made in Israel: Exploiting Palestinian land for treatment of Israeli waste.* B'tselem report. Available at: www.btselem.org/sites/default/files/publications/201712_made_in_israel_eng.pdf

Amankwaa, E. (2013). Livelihoods in risk: Exploring health and environmental implications of e-waste recycling as a livelihood strategy in Ghana. *The Journal of Modern African Studies*, 51(4), 551–575.

Amoyaw-Osei, Y., Agyekum, O. O., Pwamang, J. A., Müller, E., Fasko, R., & Schluep, M. (2011). *Ghana e-waste country assessment.* SBC e-waste Africa project. Available at: www.basel.int/Portals/4/Basel%20Convention/docs/eWaste/E-wasteAssessmentGhana.pdf

Amundsen, I., Giacaman, G., & Khan, M. H. (Eds.). (2004). *State formation in Palestine: Viability and governance during a social transformation.* New York: Routledge.

Andrade, D. F., Romanelli, J. P., & Pereira-Filho, E. R. (2019). Past and emerging topics related to electronic waste management: Top countries, trends, and perspectives. *Environmental Science and Pollution Research*, 26(17), 17135–17151.

Anyinam, C. (1994). Spatial implications of structural adjustment programmes in Ghana. *Tijdschrift Voor Economische En Sociale Geografie*, 85(5), 446–460.

ARIJ (Applied Research Institute – Jerusalem). (2009). Idhna town profile. Available at: http://vprofile.arij.org/hebron/pdfs/Idhna.pdf

(2012). The impacts of electronic waste disposal on the environmental and public health in the occupied Palestinian territory: A case study from Idhna, Hebron Governorate. Available at: www.arij.org/files/admin/latestnews/The%20impacts%20of%20electronic%20waste%20disposal%20in%20the%20occupied%20Palestinian%20territory.pdf

(2015). *Electronic waste in the town of Idhna, Hebron Governorate.* Policy paper prepared for the Palestinian Environmental Quality Authority. Hebron.

Atasu, A., & Subramanian, R. (2012). Extended producer responsibility for e-waste: Individual or collective producer responsibility? *Production and Operations Management*, 21(6), 1042–1059.

Auclair, B. R., Ormes, G. C., Smith, D. A., & Tourtillott, N. J. (2020). *Mapping informal e-waste hubs from academic and news literature.* Interactive Qualifying Project (IQP). Worcester Polytechnic Institute, Worcester, MA.

Awasthi, A. K., Zeng, X., & Li, J. (2016). Environmental pollution of electronic waste recycling in India: A critical review. *Environmental Pollution*, 211, 259–270.

Baghurst, P. A., McMichael, A. J., Wigg, N. R., Vimpani, G. V., Robertson, E. F., Roberts, R. J., & Tong, S. L. (1993). Environmental exposure to lead and children's intelligence at the age of seven years: The Port Pirie cohort study. *Obstetrical & Gynecological Survey*, 48(4), 230–231.

Bailony, M. R., Hararah, M. K., Salhab, A. R., Ghannam, I., Abdeen, Z., & Ghannam, J. (2011). Cancer registration and healthcare access in West Bank, Palestine: A GIS analysis of childhood cancer, 1998–2007. *International Journal of Cancer*, 129(5), 1180–1189.

Balde, C. P., Forti, V., Gray, V., Kuehr, R., & Stegmann, P. (2017). *The global e-waste monitor: 2017.* Bonn/Geneva/Vienna: United Nations University (UNU), International Telecommunication Union (ITU) & International Solid Waste Association (ISWA).

Basole, A. (2016). Informality and flexible specialization: Apprenticeships and knowledge spillovers in an Indian silk weaving cluster. *Development and Change*, 47(1), 157–187.

Baviskar, A. (2011). Cows, cars and cycle-rickshaws: Bourgeois environmentalism and the battle for Delhi's streets. In A. Baviskar & R. Ray (Eds.), *Elite and everyman: The cultural politics of the Indian middle classes.* New Delhi, India: Routledge.

Bellinger, D., Leviton, A., Allred, E., & Rabinowitz, M. (1994). Pre-and postnatal lead exposure and behavior problems in school-aged children. *Environmental Research*, 66(1), 12–30.

Bellinger, D. C., Stiles, K. M., & Needleman, H. L. (1992). Low-level lead exposure, intelligence and academic achievement: A long-term follow-up study. *Pediatrics*, 90(6), 855–861.

Bernard, S. (2015). North–south trade in reusable goods: Green design meets illegal shipments of waste. *Journal of Environmental Economics and Management*, 69, 22–35.

Bi, X., Feng, X., Yang, Y., Li, X., Shin, G. P., Li, F., … & Fu, Z. (2009). Allocation and source attribution of lead and cadmium in maize (*Zea mays* L.) impacted by smelting emissions. *Environmental Pollution*, 157 (3), 834–839.

Borthakur, A., & Govind, M. (2017). How well are we managing E-waste in India: Evidences from the city of Bangalore. *Energy, Ecology and Environment*, 2(4), 225–235.

Borthakur, A., & Singh, P. (2020). Mapping the emergence of research activities on e-waste: A scientometric analysis and an in-depth review. In M. N. V. Prasad, M. Vithanage, & A. Borthakur (Eds.), *Handbook of electronic waste management: International best practices and case studies*. Cambridge, MA: Butterworth-Heinemann.

Botes, L., & Van Rensburg, D. (2000). Community participation in development: Nine plagues and twelve commandments. *Community Development Journal*, 35(1), 41–58.

Braunerhjelm, P., & Feldman, M. (2008). Cluster genesis: Technology-based industrial development. *Economic Geography*, 84(2), 245–246.

Breivik, K., Armitage, J. M., Wania, F., & Jones, K. C. (2014). Tracking the global generation and exports of e-waste. Do existing estimates add up? *Environmental Science & Technology*, 48(15), 8735–8743.

Brett, E. A. (2003). Participation and accountability in development management. *The Journal of Development Studies*, 40(2), 1–29.

Brigden, K., Labunska, I., Santillo, D., & Allsopp, D. (2005). *Recycling of electronic wastes in China and India: Workplace and environmental contamination*. Amsterdam, Netherlands: Greenpeace International.

Brooks, A. (2013). Stretching global production networks: The international second-hand clothing trade. *Geoforum*, 44, 10–22.

Brydon-Miller, M., Greenwood, D., & Maguire, P. (2003). Why action research? *Action Research*, 1(1), 9–28.

B'Tselem (2017). Made in Israel: Exploiting Palestinian Land for Treatment of Israeli Waste. Available at: www.btselem.org/publications/summaries/201712_made_in_israel

Cai, Q. Y., Mo, C. H., Li, Y. H., Zeng, Q. Y., Katsoyiannis, A., Wu, Q. T., & Férard, J. F. (2007). Occurrence and assessment of polycyclic aromatic hydrocarbons in soils from vegetable fields of the Pearl River Delta, South China. *Chemosphere*, 68(1), 159–168.

Cameron, J., & Gibson, K. (2005). Participatory action research in a post-structuralist vein. *Geoforum*, 36(3), 315–331.

Cazabon, D., Fobil, J. N., Essegbey, G., & Basu, N. (2017). Structured identification of response options to address environmental health risks at the Agbogbloshie electronic waste site. *Integrated Environmental Assessment and Management*, 13(6), 980–991.

Chambers, R. (2008). PRA, PLA and pluralism: Practice and theory. In P. Reason & H. Bradbury (Eds.), *The Sage handbook of action research: Participatory inquiry and practice*. London, UK: Sage Publications.

Chan, J. K. Y., & Wong, M. H. (2013). A review of environmental fate, body burdens, and human health risk assessment of PCDD/Fs at two typical electronic waste recycling sites in China. *Science of the Total Environment*, 463, 1111–1123.

Chen, A., Dietrich, K. N., Huo, X., & Ho, S. M. (2011). Developmental neurotoxicants in e-waste: An emerging health concern. *Environmental Health Perspectives*, 119(4), 431–438.

Chen, D., Faibil, D., & Agyemang, M. (2020). Evaluating critical barriers and pathways to implementation of e-waste formalization management systems in Ghana: A hybrid BWM and fuzzy TOPSIS approach. *Environmental Science and Pollution Research*, 27(35), 44561–44584.

Chen, F., Li, X., Ma, J., Qu, J., Yang, Y., & Zhang, S. (2019). Remediation of soil co-contaminated with decabromodiphenyl ether (BDE-209) and copper by enhanced electrokinetics-persulfate process. *Journal of Hazardous Materials*, 369, 448–455.

Chen, W. Q., & Graedel, T. E. (2012). Anthropogenic cycles of the elements: A critical review. *Environmental Science & Technology*, 46(16), 8574–8586.

Chi, X., Streicher-Porte, M., Wang, M. Y., & Reuter, M. A. (2011). Informal electronic waste recycling: A sector review with special focus on China. *Waste Management*, 31(4), 731–742.

Chigbo, C., Batty, L., & Bartlett, R. (2013). Interactions of copper and pyrene on phytoremediation potential of *Brassica juncea* in copper–pyrene co-contaminated soil. *Chemosphere*, 90(10), 2542–2548.

Ciesielski, T., Weuve, J., Bellinger, D. C., Schwartz, J., Lanphear, B., & Wright, R. O. (2012). Cadmium exposure and neurodevelopmental outcomes in US children. *Environmental Health Perspectives*, 120(5), 758–763.

Ciocoiu, C. N. (2012). The role of informal sector within WEEE management systems: A Romanian perspective. *Theoretical and Empirical Researches in Urban Management*, 7(1), 27–38.

Ciocoiu, C. N., Hincu, D. L., Dobrea, R. C., & Clesca, S. E. (2011). Performance of WEEE management system in Romania. *Recent Researches in Urban Sustainability and Green Development*, 181, 178–182.

Clapp, J. (2001). *Toxic exports: The transfer of hazardous waste from rich to poor countries*. Ithaca, NY: Cornell University Press.

Cleaver, F. (1999). Paradoxes of participation: Questioning participatory approaches to development. *Journal of International Development*, 11 (4), 597–612.

Coe, N. M., & Yeung, H. W. C. (2019). Global production networks: Mapping recent conceptual developments. *Journal of Economic Geography*, 19(4), 775–801.

Cohen, S. (2010). Environmentalism deferred: Nationalisms and Israeli/ Palestinian imaginaries. In D. K. Davis & E. Burke (Eds.), *Environmental histories of the Middle East and North Africa*. Athens, OH: Ohio University Press.

Coletto, D., & Bisschop, L. (2017). Waste pickers in the informal economy of the global South: Included or excluded? *International Journal of Sociology and Social Policy*, 37(5–6), 280–294.

Cook, I. (2004). Follow the thing: Papaya. *Antipode*, 36(4), 642–664.

Cooke, B., & Kothari, U. (2001a). The social psychological limits of participation? In B. Cooke & U. Kothari (Eds.), *Participation: The new tyranny?* London, UK: Zed Books.

(2001b). The case for participation as tyranny. In B. Cooke & U. Kothari (Eds.), *Participation: The new tyranny?* London, UK: Zed Books.

Cornwall, A. (2003). Whose voices? Whose choices? Reflections on gender and participatory development. *World Development*, 31(8), 1325–1342.

Cornwall, A., & Jewkes, R. (1995). What is participatory research? *Social Science Medicine* 41(12), 1667–1676.

Corwin, J. E. (2018). "Nothing is useless in nature": Delhi's repair economies and value-creation in an electronics "waste" sector. *Environment and Planning A: Economy and Space*, 50(1), 14–30.

Crang, M., Hughes, A., Gregson, N., Norris, L., & Ahamed, F. (2013). Rethinking governance and value in commodity chains through global recycling networks. *Transactions of the Institute of British Geographers*, 38(1), 12–24.

Damrongsiri, S., Vassanadumrongdee, S., & Tanwattana, P. (2016). Heavy metal contamination characteristic of soil in WEEE (waste electrical and electronic equipment) dismantling community: A case study of Bangkok, Thailand. *Environmental Science and Pollution Research*, 23(17), 17026–17034.

Davis, J. M. (2013). The informal e-waste sector of Israel-Palestine: Contexts, problems and ways forwards. (Master's Thesis), Ben Gurion University, Beersheva, Israel.

(2017). Defining and Equitable Israeli-Palestinian E-waste Economy. (Doctoral Dissertations), Memorial University of Newfoundland, St. John's, Canada.

(2020). A model to rapidly assess informal e-waste systems. *Waste Management & Research*, 39(1), 101–107.

Davis, J. M., Akese, G., & Garb, Y. (2018). Beyond the Pollution Haven Hypothesis: Where and why do e-waste hubs emerge and what does this mean for policies and interventions? *Geoforum*, 98, 36–45.

Davis, J. M., & Garb, Y. (2015). A model for partnering with the informal e-waste industry: rationale, principles and a case study. *Resources, Conservation and Recycling*, 105, 73–83.

(2016). Cycles of waste and value: The Israeli-Palestinian e-waste system. Techniques & Culture. *Fixing the World: Excess, Leftover, and Innovation*, 390–403.

(2017). Participatory shaping of community futures in e-waste processing hubs: Complexity, conflict, and stewarded convergence in a Palestinian context. *Development Policy Review*, 37(1), 67–89:

(2018a). Quantifying flows and economies of informal e-waste hubs: Learning from the Israeli-Palestinian e-waste sector. *The Geographical Journal*, 185(1), 82–95.

(2018b). A strong spatial association between e-waste burn sites and childhood lymphoma in the West Bank, *Palestine. International Journal of Cancer*, 144(3), 470–475.

(2019). Extended responsibility or continued dis/articulation? Critical perspectives on electronic waste policies from the Israeli-Palestinian case. *Environment and Planning E: Nature and Space*, 2(2), 368–389.

(2020). Toward active community environmental policing: Potentials and limits of a catalytic model. *Environmental Management*, 65(3), 385–398.

Davis, J. M., Akese, G., & Garb, Y. (2019). Beyond the pollution haven hypothesis: Where and why do e-waste hubs emerge and what does this mean for policies and interventions? *Geoforum*, 98, 36–45.

de Oliveira, J. A. P. (Ed.). (2008). *Upgrading clusters and small enterprises in developing countries: Environmental, labor, innovation and social issues*. Surrey, UK: Ashgate Publishing.

Dias, S. M. (2016). Waste pickers and cities. *Environment and Urbanization*, 28(2), 375–390.

Dill, B. (2009). The paradoxes of community-based participation in Dar es Salaam. *Development and Change*, 40(4), 717–743.

Doron, A., & Jeffery, R. (July 2018). India's unofficial recycling bin: The city where electronics go to die. *The Guardian*. Available at: www.theguardian.com/cities/2018/jul/09/indias-unofficial-recycling-bin-the-city-where-electronics-go-to-die-moradabad

Efthymiou, L., Mavragani, A., & Tsagarakis, K. P. (2016). Quantifying the effect of macroeconomic and social factors on illegal e-waste trade. *International Journal of Environmental Research and Public Health*, 13(8), 789.

Eversole, R. (2003). Managing the pitfalls of participatory development: Some insight from Australia. *World Development*, 31(5), 781–795.

Forti, V., Balde, C. P., Kuehr, R., & Bel, G. (2020). The global e-waste monitor 2020: Quantities, flows, and the circular economy potential. United Nations University (UNU)/United Nations Institute for Training and Research (UNITAR) – co-hosted SCYCLE Programme, International Telecommunication Union (ITU) & International Solid Waste Association (ISWA).

Francis, P. (2002). Participatory development at the World Bank: The primacy of process. In B. Cooke & U. Kothari (Eds.), *Participation: The new tyranny?* London, UK: Zed Books.

Friedlander, L. R., Weisbrod, N., & Garb, Y. J. (2019). Climatic and soil-mineralogical controls on the mobility of trace metal contamination released by informal electronic waste (e-waste) processing. *Chemosphere*, 232, 130–139.

Frontline. (June 2009). Ghana: Digital dumping ground. Available at: www.pbs.org/frontlineworld/stories/ghana804/slideshow/slideshow.html

Fu, J., Zhou, Q., Liu, J., Liu, W., Wang, T., Zhang, Q., & Jiang, G. (2008). High levels of heavy metals in rice (*Oryzasativa* L.) from a typical e-waste recycling area in southeast China and its potential risk to human health. *Chemosphere*, 71(7), 1269–1275.

Fujimori, T., & Takigami, H. (2014). Pollution distribution of heavy metals in surface soil at an informal electronic-waste recycling site. *Environmental Geochemistry and Health*, 36(1), 159–168.

Funder, M. (2010). The social shaping of participatory spaces: Evidence from community development in Southern Thailand. *The Journal of Development Studies*, 46(10), 1708–1728.

Furniss, J. (2015). Alternative framings of transnational waste flows: Reflections based on the Egypt–China PET plastic trade. *Area*, 47(1), 24–30.

Galison, P. (1997). *Image and logic: A material culture of microphysics*. University of Chicago Press.

Gao, Y., Ge, L., Shi, S., Sun, Y., Liu, M., Wang, B., … & Tian, J. (2019). Global trends and future prospects of e-waste research: A bibliometric analysis. *Environmental Science and Pollution Research*, 26(17), 17809–17820.

Garb, Y. (March 2008). A Survey of Freight Traffic between Israel and the Southern West Bank and the Tarqumiya Crossing in Particular. A report funded by USAID and submitted to the Economic Cooperation Foundation.

Garb, Y., Davis, J. M. (2023). Hub-driven policy packages as a basis for e-waste reform: rationales and a case study. *Environmental Politics*, 32 (6), 1033–1053.

Geissdoerfer, M., Savaget, P., Bocken, N. M., & Hultink, E. J. (2017). The circular economy: A new sustainability paradigm? *Journal of Cleaner Production*, 143, 757–768.

Gidwani, V. (2015). The work of waste: Inside India's infra-economy. *Transactions of the Institute of British Geographers*, 40(4), 575–595.

Gidwani, V., & Reddy, R. N. (2011). The afterlives of "waste": Notes from India for a minor history of capitalist surplus. *Antipode*, 43(5), 1625–1658.

Givoni, M., Macmillen, J., Banister, D., & Feitelson, E. (2013). From policy measures to policy packages. *Transport Reviews*, 33(1), 1–20.

GLSHD (Green Land Society for Health Development). (2013). Solutions for electronic waste disposal affecting Palestinians and Israelis. Grant proposal submitted to EU Partnership for Peace Programme.

Goldstein, J. (2017). A pyrrhic victory? The limits to the successful crackdown on informal-sector plastics recycling in Wenan County, China. *Modern China*, 43(1), 3–35.

Gordon, N. (2008). *Israel's occupation*. Los Angeles: University of California Press.

Grant, K., Goldizen, F. C., Sly, P. D., Brune, M. N., Neira, M., van den Berg, M., & Norman, R. E. (2013). Health consequences of exposure to e-waste: A systematic review. *The Lancet Global Health*, 1(6), e350–e361.

Grant, R. (2006). Out of place? Global citizens in local spaces: A study of the informal settlements in the Korle Lagoon Environs in Accra, Ghana. *Urban Forum*, 17(1), 1–24.

(2019). E-waste challenges in Cape Town: Opportunity for the green economy? *Urbani izziv*, 30, 5–23.

Grant, R., & Oteng-Ababio, M. (2012). Mapping the invisible and real "African" economy: Urban e-waste circuitry. *Urban Geography*, 33(1), 1–21.

(2021). Formalising e-waste in Ghana: An emerging landscape of fragmentation and enduring barriers. *Development Southern Africa*, 38(1), 73–86.

Green, M. (2010). Making development agents: Participation as boundary object in international development. *The Journal of Development Studies*, 46(7), 1240–1263.

Greenpeace (February 2009). *Where Does E-waste End Up?* Available at: www.greenpeace.org/international/en/campaigns/detox/electronics/the-e-waste-problem/where-does-e-waste-end-up/

Gregson, N., & Crang, M. (2015). From waste to resource: The trade in wastes and global recycling economies. *Annual Review of Environment and Resources*, 40, 151–176.

Gui, L., Atasu, A., Ergun, Ö., & Toktay, L. B. (2013). Implementing extended producer responsibility legislation. *Journal of Industrial Ecology*, 17(2), 262–276.

Guijt, I., & Shah, M. K. (1998). *The myth of community: Gender issues in participatory development*. London, UK: Intermediate Technology Publications.

Gullett, B. K., Linak, W. P., Touati, A., Wasson, S. J., Gatica, S., & King, C. J. (2007). Characterization of air emissions and residual ash from open burning of electronic wastes during simulated rudimentary recycling operations. *Journal of Material Cycles and Waste Management*, 9(1), 69–79.

Gunsilius, E. (2010). Role of informal sector in solid waste management and enabling conditions for its integration. Experiences from GTZ. *Transwaste Workshop on the Informal Sector*. Available at: www.transwaste.eu/file/001441.pdf

Gutberlet, J. (2012). Informal and cooperative recycling as a poverty eradication strategy. *Geography Compass*, 6(1), 19–34.

Gutberlet, J., Carenzo, S., Kain, J. H., & Mantovani Martiniano de Azevedo, A. (2017). Waste picker organizations and their contribution to the circular economy: Two case studies from a global south perspective. *Resources*, 6(4), 52–64.

Gutberlet, J., & Carenzo, S. (2020). Waste pickers at the heart of the circular economy: A perspective of inclusive recycling from the Global South. *Worldwide Waste: Journal of Interdisciplinary Studies*, 3(1), 1–14.

Gutiérrez-Ginés, M. J., Hernández, A. J., Pérez-Leblic, M. I., Pastor, J., & Vangronsveld, J. (2014). Phytoremediation of soils co-contaminated by organic compounds and heavy metals: Bioassays with *Lupinus luteus* L. and associated endophytic bacteria. *Journal of Environmental Management*, 143, 197–207.

Haefliger, P., Mathieu-Nolf, M., Lociciro, S., Ndiaye, C., Coly, M., Diouf, A., . . . & Neira, M. (2009). Mass lead intoxication from informal used lead-acid battery recycling in Dakar, Senegal. *Environmental Health Perspectives*, 117(10), 1535–1540.

Harris, J. (2014). The messy reality of agglomeration economies in urban informality: Evidence from Nairobi's handicraft industry. *World Development*, 61, 102–113.

Hartmann, C. (2018). Waste picker livelihoods and inclusive neoliberal municipal solid waste management policies: The case of the La Chureca garbage dump site in Managua, Nicaragua. *Waste Management*, 71, 565–577.

Hayami, Y., Dikshit, A. K., & Mishra, S. N. (2006). Waste pickers and collectors in Delhi: Poverty and environment in an urban informal sector. *The Journal of Development Studies*, 42(1), 41–69.

Hayward, C., Simpson, L., & Wood, L. (2004). Still left out in the cold: Problematising participatory research and development. *Sociologia Ruralis*, 44(1), 95–108.

He, M., Shen, H., Li, Z., Wang, L., Wang, F., Zhao, K., . . . & Xu, J. (2019). Ten-year regional monitoring of soil-rice grain contamination by heavy metals with implications for target remediation and food safety. *Environmental Pollution*, 244, 431–439.

Hedley, A. J., Wong, T. W., Hui, L. L., Malisch, R., & Nelson, E. A. (2006). Breast milk dioxins in Hong Kong and Pearl River Delta. *Environmental Health Perspectives*, 114(2), 202–208.

Herod, A., Pickren, G., Rainnie, A., & McGrath Champ, S. (2014). Global destruction networks, labour and waste. *Journal of Economic Geography*, 14(2), 421–441.

Hever, S. (2010). *The Political Economy of Israel's Occupation: Repression Beyond Exploitation*. London: Pluto Press.

(2008). The return of politics in development studies I: Getting lost within the poverty agenda. *Progress in Development Studies*, 8(4), 349–358.

Hickey, S., & Mohan, G. (2004). *Participation from tyranny to transformation? Exploring new approaches to participation in development.* London, UK: Zed books.

(2005). Relocating participation within a radical politics of development. *Development and Change*, 36(2), 237–262.

Hicks, C., Dietmar, R., & Eugster, M. (2005). The recycling and disposal of electrical and electronic waste in China: Legislative and market responses. *Environmental Impact Assessment Review*, 25(5), 459–471.

Hilal, J., Kafri, S. A., & Kuttab, E. (2008). Unprotected employment in the West Bank and Gaza Strip: A gender equality and workers' rights perspective. International Labour Organization. Available at: www.ilo .org/wcmsp5/groups/public/—dgreports/—gender/documents/publica tion/wcms_097715.pdf

Howlett, M., & Rayner, J. (2007). Design principles for policy mixes: Cohesion and coherence in 'new governance arrangements.' *Policy and Society*, 26(4), 1–18.

Huang, C. L., Bao, L. J., Luo, P., Wang, Z. Y., Li, S. M., & Zeng, E. Y. (2016). Potential health risk for residents around a typical e-waste recycling zone via inhalation of size-fractionated particle-bound heavy metals. *Journal of Hazardous Materials*, 317, 449–456.

Huang, D. Y., Liu, C. P., Li, F. B., Liu, T. X., Liu, C. S., Tao, L., & Wang, Y. (2014). Profiles, sources, and transport of polycyclic aromatic hydrocarbons in soils affected by electronic waste recycling in Longtang, South China. *Environmental Monitoring and Assessment*, 186(6), 3351–3364.

Huang, G., Su, X., Rizwan, M. S., Zhu, Y., & Hu, H. (2016). Chemical immobilization of Pb, Cu, and Cd by phosphate materials and calcium carbonate in contaminated soils. *Environmental Science and Pollution Research*, 23(16), 16845–16856.

Huang, H., Zhang, S., & Christie, P. (2011). Plant uptake and dissipation of PBDEs in the soils of electronic waste recycling sites. *Environmental Pollution*, 159(1), 238–243.

Hudson, R. (2004). Conceptualizing economies and their geographies: Spaces, flows and circuits. *Progress in Human Geography*, 28(4), 447–471.

Huo, X., Peng, L., Xu, X., Zheng, L., Qiu, B., Qi, Z., & Piao, Z. (2007). Elevated blood lead levels of children in Guiyu, an electronic waste recycling town in China. *Environmental Health Perspectives*, 115, (7), 1113–1117.

IARC (International Agency for Research on Cancer). (2011). IARC Monographs on the evaluation of carcinogenic risks to humans. Available at: http://monographs.iarc.fr/ENG/Monographs/PDFs/index.ph

Ijaz, A., Imran, A., ul Haq, M. A., Khan, Q. M., & Afzal, M. (2016). Phytoremediation: Recent advances in plant-endophytic synergistic interactions. *Plant and Soil*, 405(1–2), 179–195.

Iles, A. (2004). Mapping environmental justice in technology flows: Computer waste impacts in Asia. *Global Environmental Politics*, 4(4), 76–107.

Inverardi-Ferri, C. (2017). Variegated geographies of electronic waste: Policy mobility, heterogeneity and neoliberalism. *Area Development and Policy*, 2(3), 314–331.

(2021). Towards a cultural political economy of the illicit. *Progress in Human Geography*, 45(6), 1646–1667.

Isimekhai, K. A., Garelick, H., Watt, J., & Purchase, D. (2017). Heavy metals distribution and risk assessment in soil from an informal e-waste recycling site in Lagos State, Nigeria. *Environmental Science and Pollution Research*, 24(20), 17206–17219.

Israel, D. (July 2019). Smoke cloud from Arab villages' electronic waste fires hangs over Moshav Shekef. JewishPress.com. Available at: www.jewishpress.com/news/eye-on-palestine/palestinian-authority/smoke-cloud-from-arab-villages-electronic-waste-fires-hangs-over-moshav-shekef/2019/07/04/

Israel Tax Authority. (2005). Sale of metal waste – transfer tax liability from the seller to the buyer, amendment to the 1976 value added tax. Available at: https://taxes.gov.il/Vat/Pages/Parshanut/nohal2005.aspx

Itai, T., Otsuka, M., Asante, K. A., Muto, M., Opoku-Ankomah, Y., Ansa-Asare, O. D., & Tanabe, S. (2014). Variation and distribution of metals and metalloids in soil/ash mixtures from Agbogbloshie e-waste recycling site in Accra, Ghana. *Science of the Total Environment*, 470, 707–716.

Jun-Hui, Z., & Hang, M. I. N. (2009). Eco-toxicity and metal contamination of paddy soil in an e-wastes recycling area. *Journal of Hazardous Materials*, 165(1–3), 744–750.

Ka, W., Lau, Y., Liang, P., Man, Y. B., Chung, S. S., & Wong, M. H. (2014). Human health risk assessment based on trace metals in suspended air particulates, surface dust, and floor dust from e-waste recycling workshops in Hong Kong, China. *Environmental Science and Pollution Research International*, 21(5), 3813–3825.

Kahhat, R., & Williams, E. (2009). Product or waste? Importation and end-of-life processing of computers in Peru. *Environmental Science & Technology*, 43(15), 6010–6016.

(2010). Adoption and disposition of new and used computers in Lima, Peru. *Resources, Conservation and Recycling*, 54(8), 501–505.

(2012). Materials flow analysis of e-waste: Domestic flows and exports of used computers from the United States. *Resources, Conservation and Recycling*, 67, 67–74.

Kama, K. (2015). Circling the economy: Resource-making and marketization in EU electronic waste policy. *Area*, 47(1), 16–23.

Kaplan, J. (November 2015). Welcome to hell: Photographer documents Africa's e-waste nightmare. *Fox News*. Available at: www.foxnews.com/tech/welcome-to-hell-photographer-documents-africas-e-waste-nightmare

Kasinja, C., & Tilley, E. (2018). Formalization of informal waste pickers' cooperatives in Blantyre, Malawi: A feasibility assessment. *Sustainability*, 10(4), 1149–1166.

Kawasmi, H., & White, S. (2010). Towards a policy framework for the development of micro, small and medium-sized enterprises in the Occupied Palestine Territory: Assessment Report. International Labour Organization.

Kellenberg, D. (2010). Consumer waste, backhauling, and pollution havens. *Journal of Applied Economics*, 13(2), 283–304.

Kezar, A., & Maxey, D. (2016). The Delphi technique: An untapped approach of participatory research. *International Journal of Social Research Methodology*, 19(2), 143–160.

Khalidi, R. J., & Taghdisi-Rad, S. (2009). The economic dimensions of prolonged occupation: Continuity and change in Israeli policy towards the Palestinian economy. Available at: http://mpra.ub.uni-muenchen.de/21808/

Kirby, P. W., & Lora-Wainwright, A. (2015). Exporting harm, scavenging value: transnational circuits of e-waste between Japan, China and beyond. *Area*, 47(1), 40–47.

Koff, H. (2017). Policy coherence for development and migration: Analyzing US and EU policies through the lens of normative transformation. *Regions and Cohesion*, 7(2), 5–33.

Koff, H., & Maganda, C. (2019). Saving the baby while discarding the bathwater: The application of policy coherence for development analysis to payment for watershed services. *Madera y Bosques*, 25(3), 1–20.

Kojima, M., Yoshida, A., & Sasaki, S. (2009). Difficulties in applying extended producer responsibility policies in developing countries: Case studies in e-waste recycling in China and Thailand. *Journal of Material Cycles and Waste Management*, 11(3), 263–269.

Konadu-Agyemang, K. (2000). The best of times and the worst of times: Structural adjustment programs and uneven development in Africa: The case of Ghana. *The Professional Geographer*, 52(3), 469–483.

Kraemer-Mbula, E., & Wunsch-Vincent, S. (Eds.) (2016). *The informal economy in developing nations: Hidden engine of innovation?* Cambridge, UK: Cambridge University Press.

Kumar, A., Samadder, S. R., Kumar, N., & Singh, C. (2018). Estimation of the generation rate of different types of plastic wastes and possible revenue recovery from informal recycling. *Waste Management*, 79, 781–790.

Kunz, N., Atasu, A., Mayers, K., & Van Wassenhove, L. (2014). Extended producer responsibility: Stakeholder concerns and future developments. White Paper, INSEAD Social Innovation Centre, Fontainebleau.

Kunz, N., Mayers, K., & Van Wassenhove, L. N. (2018). Stakeholder views on extended producer responsibility and the circular economy. *California Management Review*, 60(3), 45–70.

Kuo, L. J., Cade, S. E., Cullinan, V., & Schultz, I. R. (2019). Polybrominated diphenyl ethers (PBDEs) in plasma from E-waste recyclers, outdoor and indoor workers in the Puget Sound, WA region. *Chemosphere*, 219, 209–216.

Kyamusugulwa, P. (2013). Participatory development and reconstruction: A literature review. *Third World Quarterly*, 34(7), 1265–1278.

Kyamusugulwa, P. M., & Hilhorst, D. (2015). Power holders and social dynamics of participatory development and reconstruction: Cases from the Democratic Republic of Congo. *World Development*, 70, 249–259.

Laha, S. (2014). Informality in e-waste processing: An analysis of the Indian experience. *Competition & Change*, 18(4), 309–326.

Lange, S. (2008). The depoliticisation of development and the democratisation of politics in Tanzania: Parallel structures as obstacles to delivering services to the poor. *The Journal of Development Studies*, 44(8), 1122–1144.

Langrová, V. (2002). Comparative analysis of EPR programmes for small consumer batteries: Case study of the Netherlands, Switzerland and Sweden. IIIEE Report. Available at: https://portal.research.lu.se/en/publications/comparative-analysis-of-epr-programmes-for-small-consumer-batteri

Lawhon, M. (2012a). Relational power in the governance of a South African e-waste transition. *Environment and Planning A*, 44(4), 954–971.

(2012b). Contesting power, trust and legitimacy in the South African e-waste transition. *Policy Sciences*, 45(1), 69–86.

(2013). Dumping ground or country-in-transition? Discourses of e-waste in South Africa. *Environment and Planning C: Government and Policy*, 31(4), 700–715.

Leal, P. A. (2007). Participation: The ascendancy of a buzzword in the neo-liberal era. *Development in Practice*, 17(4–5), 539–548.

Lee, D., Offenhuber, D., Duarte, F., Biderman, F., & Ratti, C. (2018). Monitour: Tracking global routes of electronic waste. *Waste Management*, 72, 362–370.

Lepawsky, J. (2012). Legal geographies of e-waste legislation in Canada and the US: Jurisdiction, responsibility and the taboo of production. *Geoforum*, 43(6), 1194–1206.

(2015). Are we living in a post-Basel world? *Area*, 47(1), 7–15.

(2018). *Reassembling rubbish: Worlding electronic waste*. Cambridge, MA: MIT Press.

(2020). Planet of fixers? Mapping the middle grounds of independent and do-it-yourself information and communication technology maintenance and repair. *Geo: Geography and Environment*, 7(1), 1–17.

Lepawsky, J., Araujo, E., Davis, J. M., & Kahhat, R. (2017). Best of two worlds? Towards ethical electronics repair, reuse, repurposing and recycling? *Geoforum*, 81, 87–99.

Lepawsky, J., & Billah, M. (2011). Making chains that (un)make things: Waste–value relations and the Bangladeshi rubbish electronics industry. *Geografiska Annaler: Series B, Human Geography*, 93(2), 121–139.

Lepawsky, J., & Mather, C. (2011). From beginnings and endings to boundaries and edges: Rethinking circulation and exchange through electronic waste. *Area*, 43(3), 242–249.

Lepawsky, J., & McNabb, C. (2010). Mapping international flows of electronic waste. *The Canadian Geographer*, 54(2), 177–195.

Leung, A., Cai, Z. W., & Wong, M. H. (2006). Environmental contamination from electronic waste recycling at Guiyu, southeast China. *Journal of Material Cycles and Waste Management*, 8(1), 21–33.

Leung, A. O., Cheung, K. C., & Wong, M. H. (2015). Spatial distribution of polycyclic aromatic hydrocarbons in soil, sediment, and combusted residue at an e-waste processing site in southeast China. *Environmental Science and Pollution Research*, 22(12), 8786–8801.

Leung, A. O., Duzgoren-Aydin, N. S., Cheung, K. C., & Wong, M. H. (2008). Heavy metals concentrations of surface dust from e-waste recycling and its human health implications in southeast China. *Environmental Science & Technology*, 42(7), 2674–2680.

Levy, G., & Levac, A. (December 2018). Palestinian's penalty for air pollution: 20 bullets fired by Israeli police. *Haaretz*. Available at: www.haaretz.com/israel-news/.premium-palestinian-s-penalty-for-air-pollution-20-bullets-fired-by-israeli-police-1.6787334

Li, B., Du, H., Bao, J., Higano, Y., & Li, Y. (February 2011). Policy on e-waste in China: Case study of Guiyu Town, Guangdong Province. In Computer Distributed Control and Intelligent Environmental Monitoring (CDCIEM).

Li, H., Bai, J., Li, Y., Cheng, H., Zeng, E. Y., & You, J. (2011). Short-range transport of contaminants released from e-waste recycling site in South China. *Journal of Environmental Monitoring*, 13(4), 836–843.

Li, J., Duan, H., & Shi, P. (2011). Heavy metal contamination of surface soil in electronic waste dismantling area: Site investigation and source apportionment analysis. *Waste Management Research*, 29(7), 727–738.

Liboiron, M. (January 2014). Against awareness, for scale: Garbage is infrastructure, not behavior. *Discard Studies Blog*. Available at: https://discardstudies.com/2014/01/23/against-awareness-for-scale-garbage-is-infrastructure-not-behavior/

Little, P. (November 2016). On electronic pyropolitics and Pure Earth friction in Agbogbloshie. *Toxic News*. Available at: https://toxicnews.org/2016/11/08/on-electronic-pyropolitics-and-pure-earth-friction-in-agbogbloshie/

(2018). Bodies, toxins, and e-waste labour interventions in Ghana: Toward a toxic postcolonial corporality? *Revista de Antropologia Iberoamericana*, 14(1), 51–71.

Liu, H. (2010). Analysis of the impact of globalization on the Palestinian economy. *International Journal of Business and Management*, 5(7), 204–207.

Liu, L., Hu, L., Tang, J., Li, Y., Zhang, Q., & Chen, X. (2012). Food safety assessment of planting patterns of four vegetable-type crops grown in soil contaminated by electronic waste activities. *Journal of Environmental Management*, 93(1), 22–30.

Lopez, B. N., Man, Y. B., Zhao, Y. G., Zheng, J. S., Leung, A. O. W., Yao, J., & Wong, M. H. (2011). Major pollutants in soils of abandoned agricultural land contaminated by e-waste activities in Hong Kong. *Archives of Environmental Contamination and Toxicology*, 61(1), 101–114.

Lora-Wainwright, A. (2016). The trouble of connection: E-waste in China between state regulation, development regimes and global capitalism. In K. Harper, I. Vaccaro, S. Murray (Eds.), *The anthropology of disconnection: The political ecology of postindustrial regimes*. New York, NY: Bergahn.

Lu, C., Zhang, L., Zhong, Y., Ren, W., Tobias, M., Mu, Z., . . . & Xue, B. (2015). An overview of e-waste management in China. *Journal of Material Cycles and Waste Management*, 17(1), 1–12.

Lucier, C. A., & Gareau, B. J. (2015). From waste to resources? Interrogating 'race to the bottom' in the global environmental governance of the hazardous waste trade. *Journal of World-Systems Research*, 21(2), 495–520.

Luo, C., Liu, C., Wang, Y., Liu, X., Li, F., Zhang, G., & Li, X. (2011). Heavy metal contamination in soils and vegetables near an e-waste processing site, south China. *Journal of Hazardous Materials*, 186(1), 481–490.

Luo, C., Shen, Z., Lou, L., & Li, X. (2006). EDDS and EDTA-enhanced phytoextraction of metals from artificially contaminated soil and residual effects of chelant compounds. *Environmental Pollution*, 144 (3), 862–871.

Luo, C., Shen, Z., & Li, X. (2005). Enhanced phytoextraction of Cu, Pb, Zn and Cd with EDTA and EDDS. *Chemosphere*, 59(1), 1–11.

Ma, J., Zhang, Q., Chen, F., Zhu, Q., Wang, Y., & Liu, G. (2020). Remediation of PBDEs-metal co-contaminated soil by the combination of metal stabilization, persulfate oxidation and bioremediation. *Chemosphere*, 252, 1–8.

Maliszewska-Kordybach, B. (1996). Polycyclic aromatic hydrocarbons in agricultural soils in Poland: preliminary proposals for criteria to evaluate the level of soil contamination. *Applied Geochemistry*, 11(1–2), 121–127.

Maliszewska-Kordybach, B., Smreczak, B., & Klimkowicz-Pawlas, A. (2009). Concentrations, sources, and spatial distribution of individual polycyclic aromatic hydrocarbons (PAHs) in agricultural soils in the Eastern part of the EU: Poland as a case study. *Science of the Total Environment*, 407(12), 3746–3753.

Manomaivibool, P. (2009). Extended producer responsibility in a non-OECD context: The management of waste electrical and electronic equipment in India. *Resources, Conservation and Recycling*, 53(3), 136–144.

Maguire, P. (1987). *Doing participatory research: A feminist approach*. Amherst: University of Massachusetts.

Manhart, A. (2010). International cooperation for metal recycling from waste electrical and electronic equipment. *Journal of Industrial Ecology*, 15(1), 13–30.

Manhart, A., Osibanjo, O., Aderinto, A., & Prakash, S. (2011). Informal e-waste management in Lagos, Nigeria: Socio-economic impacts and feasibility of international recycling co-operations. UNEP SBC E-waste Africa Project Report.

Manomaivibool, P., & Vassanadumrongdee, S. (2011). Extended producer responsibility in Thailand. *Journal of Industrial Ecology*, 15(2), 185–205.

Mansuri, G., & Rao, V. (2012). Localizing development: Does participation work? World Bank Policy Research Report. Washington, DC: World Bank Group. Available at: http://documents.worldbank.org/curated/en/461701468150309553/Localizing-development-does-participation-work

Mao, S., Gu, W., Bai, J., Dong, B., Huang, Q., Zhao, J., ... & Wang, J. (2020). Migration of heavy metal in electronic waste plastics during simulated recycling on a laboratory scale. *Chemosphere*, 245, 1–8.

Marshall, A. (1920). *Principles of economics*. London, UK: Macmillan.

MAS (The Palestine Economic Policy Research Institute). (April 2010). Overview of the Palestinian economy. Available at: www.pic-palestine.ps/userfiles/file/pdfs/invest_in_palestine_en.pdf

Massar Associates. (2003). Palestinian informal enterprises: Key characteristics, constraints to growth, and development opportunities. A report for USAID.

Mayoux, L. (1995). Beyond naivety: Women, gender inequality and participatory development. *Development and Change*, 26(2), 235–258.

McArdle, K. L. (2008). Getting in, getting on, getting out: On working with second-person inquiry groups. In P. Reason, & H. Bradbury (Eds.), *The Sage Handbook of action research: Participatory inquiry and practice*. London, UK: Sage Publications.

McCormick, D., & Oyelaran-Oyeyinka, B. (2007). Introduction: Clusters and innovation systems in Africa. In Oyebanji Oyelaran-Oyeyinka, & Dorothy McCormick (Eds.), *Industrial cluster and innovation systems in Africa: Institutions, markets and policy*. Tokyo, Japan: UN University Press.

McDonald, T. A. (2002). A perspective on the potential health risks of PBDEs. *Chemosphere*, 46(5), 745–755.

McElvaney, J. (February 2014). Ghana's e-waste magnet. Aljazeera. Available at: www.aljazeera.com/indepth/inpictures/2014/01/pictures-ghana-e-waste-mecca-201413010474097 5223.html

McGrath-Champ, S., Rainnie, A., Pickren, G., & Herod, A. (2015). Global destruction networks, the labour process and employment relations. *Journal of Industrial Relations*, 57(4), 624–640.

Meagher, K. (2007). Manufacturing disorder: Liberalization, informal enterprise and economic 'ungovernance' in African small firm clusters. *Development and Change*, 38(3), 473–503.

Medina, M. (2007). Waste picker cooperatives in developing countries. In M. Chen, R. Jhabvala, R. Kanbur, & C. Richards (Eds.), *Membership-based organizations of the poor*. London, UK: Routledge.

Miaari, S., & Sauer, R. (2011). The labor market costs of conflict: Closures, foreign workers, and Palestinian employment and earnings. *Review of Economics of the Household*, 9(1), 129–148.

Miaari, S., Zussman, A., & Zussman, N. (2014). Employment restrictions and political violence in the Israeli–Palestinian conflict. *Journal of Economic Behavior & Organization*, 101, 24–44.

Michener, V. (1998). The participatory approach: Contradiction and co-option in Burkina Faso. *World Development*, 26(12), 2105–2118.

Millington, N., & Lawhon, M. (2019). Geographies of waste: Conceptual vectors from the Global South. *Progress in Human Geography*, 43(6), 1044–1063.

Milojkovic, J., & Litovski, V. (2005). Concepts of computer take-back for sustainable end-of-life. *Working and Living Environment Protection*, 2 (5), 363–372.

Minter, A. (January 2016). The burning truth behind an e-waste dump in Africa. *Smithsonian Magazine*. Available at: www.smithsonianmag .com/science-nature/burning-truth-behind-e-waste-dump-africa-180957597/.

Möckel, C., Breivik, K., Nøst, T. H., Sankoh, A., Jones, K. C., & Sweetman, A. (2020). Soil pollution at a major West African e-waste recycling site: Contamination pathways and implications for potential mitigation strategies. *Environment International*, 137, 1–12.

Mohan, G., & Stokke, K. (2000). Participatory development and empowerment: The dangers of localism. *Third World Quarterly*, 21(2), 247–268.

Mosse, D. (2001). People's knowledge', participation and patronage: Operations and representations in rural development. In B. Cooke, & U. Kothari (Eds.), *Participation: The new tyranny?* London, UK: Zed Books.

Mottaleb, K. A., & Sonobe, T. (2013). The development process of rural informal industries in developing countries: The case of Bangladesh. *The Journal of Developing Areas*, 47(2), 229–249.

Mudge, S. M., Pfaffhuber, K. A., Fobil, J. N., Bouman, E. A., Uggerud, H. T., & Thorne, R. J. (2019). Using elemental analyses and multivariate statistics to identify the off-site dispersion from informal e-waste processing. *Environmental Science: Processes & Impacts*, 21(12), 2042–2057.

Nagajyoti, P. C., Lee, K. D., & Sreekanth, T. V. M. (2010). Heavy metals, occurrence and toxicity for plants: A review. *Environmental Chemistry Letters*, 8(3), 199–216.

Naqib, F. M. (2003). Economic aspects of the Palestinian—Israeli conflict: The collapse of the Oslo Accord. *Journal of International Development: The Journal of the Development Studies Association*, 15(4), 499–512.

Newman, D. (2009). In the name of security: In the name of peace–environmental schizophrenia and the security discourse in Israel/Palestine. In H. G. Brauch, U. O. Spring, C. Mesjasz, P. Kameri-Mbote, N. C. Behera, B. Chourou, & H. Krummenacher (Eds.), *Facing global environmental change*. Berlin, Germany: Springer.

Ni, H. G., & Zeng, E. Y. (2009). Law enforcement and global collaboration are the keys to containing e-waste tsunami in China. *Environmental Science & Technology*, 43(11), 3991–3994.

Ni, K., Lu, Y., Wang, T., Shi, Y., Kannan, K., Xu, L., ... & Liu, S. (2013). Polybrominated diphenyl ethers (PBDEs) in China: Policies and recommendations for sound management of plastics from electronic wastes. *Journal of Environmental Management*, 115, 114–123.

Nixon, R. (2011). *Slow violence and the environmentalism of the poor*. Cambridge, MA: Harvard University Press.

Nnorom, I. C., & Osibinjo, O. (2008a). Electronic waste (e-waste): Material flows and management practice in Nigeria. *Waste Management*, 28(8), 1472–1479.

Nnorom, I. C., & Osibanjo, O. (2008b). Overview of electronic waste (e-waste) management practices and legislations, and their poor applications in the developing countries. *Resources, Conservation and Recycling*, 52(6), 843–858.

OECD. (2001). Extended producer responsibility: A guidance manual for governments. Organisation for Economic Co-operation and Development. Available at: www.oecd-ilibrary.org/environment/extended-producer-responsibility_9789264189867-en

Ogungbuyi, O., Nnorom, I. C., Osibanjo, O., & Schluep, M. (2012). E-waste country assessment Nigeria: A report of e-waste Africa project of the Secretariat of Basel Convention. Available at: www.basel.int/Portals/4/Basel%20Convention/docs/eWaste/EwasteAfrica_Nigeria-Assessment.pdf

Oguri, T., Suzuki, G., Matsukami, H., Uchida, N., Tue, N. M., Viet, P. H., ... & Takigami, H. (2018). Exposure assessment of heavy metals in an e-waste processing area in northern Vietnam. *Science of the Total Environment*, 621, 1115–1123.

Ohajinwa, C. M., Van Bodegom, P. M., Osibanjo, O., Xie, Q., Chen, J., Vijver, M. G., & Peijnenburg, W. J. (2019). Health risks of polybrominated diphenyl ethers (PBDEs) and metals at informal electronic waste recycling sites. *International Journal of Environmental Research and Public Health*, 16(6), 906–925.

Ohajinwa, C. M., Van Bodegom, P. M., Xie, Q., Chen, J., Vijver, M. G., Osibanjo, O. O., & Peijnenburg, W. J. (2019). Hydrophobic organic pollutants in soils and dusts at electronic waste recycling sites:

Occurrence and possible impacts of polybrominated diphenyl ethers. *International Journal of Environmental Research and Public Health*, 16 (3), 360–378.

Olafisoye, O. B., Adefioye, T., & Osibote, O. A. (2013). Heavy metals contamination of water, soil, and plants around an electronic waste dumpsite. *Polish Journal of Environmental Studies*, 22(5), 1431–1439.

Oliveira, C. R., Bernardes, A. M., & Gerbase, A. E. (2012). Collection and recycling of electronic scrap: A worldwide overview and comparison with the Brazilian situation. *Waste Management*, 32(8), 1592–1610.

Osibanjo, O., & Nnorom, I. C. (2007). The challenge of electronic waste (e-waste) management in developing countries. *Waste Management & Research*, 25(6), 489–501.

Oteng-Ababio, M., Owusu-Sekyere, E., & Amoah, S. T. (2017). Thinking globally, acting locally: Formalizing informal solid waste management practices in Ghana. *Journal of Developing Societies*, 33 (1), 75–98.

Oteng-Ababio, M., Owusu, G., & Chama, M. (2016). Intelligent enterprise: Wasting, valuing and re-valuing waste electrical and electronic equipment. *The Geographical Journal*, 182(3), 265–275.

Oteng-Ababio, M., van der Velden, M., & Taylor, M. B. (2020). Building policy coherence for sound waste electrical and electronic equipment management in a developing country. *The Journal of Environment & Development*, 29(3), 306–328.

Ouabo, R. E., Sangodoyin, A. Y., & Ogundiran, M. B. (2020). Assessment of ordinary kriging and inverse distance weighting methods for modeling chromium and cadmium soil pollution in e-waste sites in Douala, Cameroon. *Journal of Health and Pollution*, 10(26), 1–20.

Palestinian Environmental Law. (1999). Title I: Definitions and General Provisions. Available at: http://faolex.fao.org/docs/texts/pal40426E.doc

Parfitt, T. (2004). The ambiguity of participation: A qualified defence of participatory development. *Third World Quarterly*, 25(3), 537–555.

Pellow, D. (2007). *Resisting global toxics: Transnational movements for environmental justice*. Cambridge, MA: The MIT Press.

Perkins, D. N., Drisse, M. N. B., Nxele, T., & Sly, P. D. (2014). E-waste: A global hazard. *Annals of Global Health*, 80(4), 286–295.

PCBS (Palestinian Central Bureau of Statistics). (2003). National accounts at current and constant prices (1994–2000). Available at: www.pcbs.gov .ps/PCBS_2012/Publications.aspx?CatId=25&scatId=297

(2013). Number of enterprises and employed persons and main economic indicators in Palestine for industrial activities, 2013. Available at: www .pcbs.gov.ps/Portals/_Rainbow/Documents/Number%20of% 20Enterprises%20and%20Employed%20Persons%20and%20Main%

20Economic%20Indicators%20in%20Palestine%20For%20Industrial %20Activities%20,%202012.htm

(2014a). Foreign trade main indicators of 2014. Available at: www.pcbs .gov.ps/Portals/_Rainbow/Documents/Main%20Indicator_E.htm

(2014b). Major national accounts variables by region for the years 1994–2014 at current prices. Available at: www.pcbs.gov.ps/Portals/_Rainbow/Documents/e-namcurr-1994–2014.htm

(2014c). Number of employed persons and compensation of paid employees for olive presses in Palestine by governorate, 2014. Available at: www.pcbs.gov.ps/Portals/_Rainbow/Documents/olive3-Annual-2014-e .htm

(2014d). Main economic indicators for olive presses activity in Palestine by governorate, 2014. Available at: www.pcbs.gov.ps/Portals/_Rainbow/Documents/olive-Main-2014-e.htm

(2014e). Number of enterprises, employed persons, and main economic indicators for the tourism enterprises in Palestine by tourism activity, 2014. Available at: www.pcbs.gov.ps/Portals/_Rainbow/Documents/TourAct-2014-E-01.htm

Plambeck, E., & Wang, Q., (2009). Effects of e-waste regulation on new product introduction. *Management Science*, 55(3), 333–347.

Platteau, J. P. (2004). Monitoring elite capture in community-driven development. *Development and Change*, 35(2), 223–246.

Ploumis, P. (January 2011). Thriving business of illegal e-waste recycling in Seelampur. *Scrap Monster*. Available at: www.scrapmonster.com/news/thriving-business-of-illegal-e-waste-recycling-in-seelampur/1/133

Portugali, J. (1993). *Implicate relations society and space in the Israeli-Palestinian conflict*. Dordrecht: Kluwer Academic Publishers.

(2013). *Implicate relations: Society and space in the Israeli-Palestinian conflict*. Dordrecht, The Netherlands: Kluwer Academic Publishers.

Pradhan, J. K., & Kumar, S. (2014). Informal e-waste recycling: Environmental risk assessment of heavy metal contamination in Mandoli industrial area, Delhi, *India. Environmental Science and Pollution Research*, 21(13), 7913–7928.

Prakash, S., Manhart, A., Agyekum, O. O., Amoyaw-Osei, Y., Schluep, M., Müller, E., & Fasko, R. (2010). Informal e-waste recycling sector in Ghana: An in-depth socio-economic study. Report from Öko-Institut (Institute for Applied Ecology).

Prakash, S., Manhart, A., Amoyaw-Osei, Y., & Agyekum, O. O. (2010). Socio-economic assessment and feasibility study on sustainable e-waste management in Ghana. Oko-Institut eV in cooperation with Ghana Environmental Protection Agency (EPA) & Green Advocacy Ghana,

Ministry of Housing, Spatial Planning and the Environment. Available at: www.oeko.de/oekodoc/1057/2010-105-en.pdf

Pratt, L. A. (2010). Decreasing dirty dumping: A reevaluation of toxic waste colonialism and the global management of transboundary hazardous waste. *State Bar of Texas Environmental Law Journal*, 41(2), 581–623.

Puckett, J., Byster, L., Westervelt, S., Gutierrez, R., Davis, S., Hussain, A., & Dutta, M. (2002). *Exporting harm: The high-tech trashing of Asia*. Seattle, WA: The Basel Action Network.

Puckett, J., Westervelt, S., Gutierrez, R., & Takamiya, Y. (2005). *The digital dump: Exporting re-use and abuse to Africa*. Seattle, WA: The Basel Action Network.

Purdey, A. F., Adhikari, G. B., Robinson, S. A., & Cox, P. W. (1994). Participatory health development in rural Nepal: Clarifying the process of community empowerment. *Health Education & Behavior*, 21(3), 329–343.

Radulovic, V. (2018). Portrayals in print: Media depictions of the informal sector's involvement in managing e-waste in India. *Sustainability*, 10(4), 966–987.

Rangoni, R., & Jager, W. (2017). Social dynamics of littering and adaptive cleaning strategies explored using agent-based modelling. *Journal of Artificial Societies and Social Simulation*, 20(2), 1–17.

Rathore, G. J. S. (2020). Circulating waste, circulating bodies? A critical review of E-waste trade. *Geoforum*, 110, 180–182.

Reason, P. (1994). Three approaches to participative inquiry. In N. K. Denzin & Y. S. Lincoln (Eds.), *Handbook of qualitative research*. Thousand Oaks, CA: Sage Publications.

Reddy, R. N. (2013). Revitalising economies of disassembly: Informal recyclers, development experts and e-waste reforms in Bangalore. *Economic and Political Weekly*, 48(13), 62–70.

(2015). Producing abjection: E-waste improvement schemes and informal recyclers of Bangalore. *Geoforum*, 62, 166–174.

(2016). Reimagining e-waste circuits: Calculation, mobile policies, and the move to urban mining in global south cities. *Urban Geography*, 37(1), 57–76.

Reynolds, K. M. (2017). Unpacking the complex nature of cooperative interactions: Case studies of Israeli–Palestinian environmental cooperation in the greater Bethlehem area. *Geojournal*, 82(4), 701–719.

Ribot, J. C. (2003). Democratic decentralization of natural resources. In N. Van de Walle, N. Ball, & V. Ramachandran (Eds.), *Beyond structural adjustment: The institutional context of African development*. New York, NY: Palgrave Macmillan.

(2004). *Waiting for democracy*. Washington, DC: World Resources Institute,.

Rigon, A. (2014). Building local governance: Participation and elite capture in slum-upgrading in Kenya. *Development and Change*, 45(2), 257–283.

Riisgaard, L., Bolwig, S., Ponte, S., Du Toit, A., Halberg, N., & Matose, F. (2010). Integrating poverty and environmental concerns into value-chain analysis: A strategic framework and practical guide. *Development Policy Review*, 28(2), 195–216.

Ritchie, M. (2000). Social capacity, sustainable development, and older people: Lessons from community-based care in Southeast Asia. *Development in Practice*, 10(5), 638–649.

Rochat, D., Rodrigues, W., & Gantenbein, A. (October 2008). India: Including the existing informal sector in a clean e-waste channel. In Proceedings of the Waste Management Conference (WasteCon2008). Durban, South Africa.

Rochman, F. F., Ashton, W. S., & Wiharjo, M. G. (2017). E-waste, money and power: Mapping electronic waste flows in Yogyakarta, Indonesia. *Environmental Development*, 24, 1–8.

Rotter, V. S., Chancerel, P., & Schill, W. P. (2011). Practicalities of individual producer responsibility under the WEEE directive: Experiences in Germany. *Waste Management & Research*, 29(9), 931–944.

Roy, S. (1998). The Palestinian economy after Oslo. *Current History*, 97 (615), 19–25.

Rutkowski, J. E., & Rutkowski, E. W. (2015). Expanding worldwide urban solid waste recycling: The Brazilian social technology in waste pickers inclusion. *Waste Management & Research*, 33(12), 1084–1093.

Sachs, N. (2006). Planning the funeral at the birth: Extended producer responsibility in the European Union and the United States. *Harvard Environmental Law Review*, 30(51), 51–98.

Samson, M. (2009). Refusing to be cast aside: Waste pickers organising around the world. *Women in Informal Employment: Globalizing and Organizing (WIEGO)*. Available at: www.wiego.org/publications/refusing-be-cast-aside-waste-pickers-organising-around-world

SBC (Secretariat of the Basel Convention). (2011). Where are WEEE in Africa? Report for the Basel Convention E-Waste Africa Programme.

Scheinberg, A., Simpson, M., & Gupt, Y. (2010). Economic aspects of the informal sector in solid waste management. Collaborative Working Group on Solid Waste Management in Low- and Middle-income Countries and GIZ - Deutsche Gesellschaft für Internationale Zusammenarbeit.

Schluep, M., Hagelüken, C., Meskers, C. E., Magalini, F., Wang, F., Müller, E., . . . & Sonnemann, G. (2009). Market potential of innovative e-waste recycling technologies in developing countries. Proceedings of the R Twin World Congress on Resource Management and Technology for Material and Energy Efficiency. Davos, Switzerland.

Schulz, Y., & Lora-Wainwright, A. (2019). In the name of circularity: Environmental improvement and business slowdown in a Chinese recycling hub. *Worldwide Waste: Journal of Interdisciplinary Studies*, 2(1), 1–13.

Sepúlveda, A., Schluep, M., Renaud, F. G., Streicher, M., Kuehr, R., Hagelüken, C., & Gerecke, A. C. (2010). A review of the environmental fate and effects of hazardous substances released from electrical and electronic equipments during recycling: Examples from China and India. *Environmental Impact Assessment Review*, 30(1), 28–41.

Sesan, T. (2014). Peeling back the layers on participatory development: Evidence from a community-based women's group in Western Kenya. *Community Development Journal*, 49(4), 603–617.

Shammai, H. E. (July 2017). Electronic waste in Israel: An overview and legal analysis of the electric and electronic waste market in Israel, and relations between the formal and informal markets. Report prepared by Adam Teva V'Din for AJEEC-NISPED.

Siegle, L. (November 2017). Reduce, reuse, reboot: Why electronic recycling must up its game. *The Guardian*. Available at: www.theguardian.com/environment/2017/nov/20/electronic-recycling-e-waste-2017-gadgets

Sihlongonyane, M. (2001). The rhetoric of the community in project management: The case of Mohlakeng township. *Development in Practice*, 11(1), 34–44.

Simpson, A. (2007). On ethnographic refusal: Indigeneity, 'voice', and colonial citizenship. *Junctures*, 9, 67–80.

Sinha-Khetriwal, D., Widmer, R., Schluep, M., Eugster, M., Wang, X., Lombard, R., & Ecroignard, L. (2006). Legislating e-waste management: Progress from various countries. *Environmental Law Network International*, 1(2), 27–36.

Smit, S., & Musango, J. (2015a). Exploring the connections between the green economy and the informal economy in South Africa. *South African Journal of Science*, 111(11–12), 1–10.

(2015b). Towards connecting the green economy with the informal economy in South Africa: A review and way forward. *Ecological Economics*, 116, 154–159.

Soetrisno, F. N., & Delgado-Saborit, J. M. (2020). Chronic exposure to heavy metals from informal e-waste recycling plants and children's

attention, executive function and academic performance. *Science of the Total Environment*, 717, 1–12.
Song, Q., & Li, J. (2014a). A systematic review of the human body burden of e-waste exposure in China. *Environment International*, 68, 82–93.
(2014b). Environmental effects of heavy metals derived from the e-waste recycling activities in China: A systematic review. *Waste Management*, 34(12), 2587–2594.
Song, Y., Wu, N., Han, J., Shen, H., Tan, Y., Ding, G., . . . & Jin, S. (2011). Levels of PCDD/Fs and DL-PCBs in selected foods and estimated dietary intake for the local residents of Luqiao and Yuhang in Zhejiang, *China. Chemosphere*, 85(3), 329–334.
Sonobe, T., Akoten, J. E., & Otsuka, K. (2011). The growth process of informal enterprises in Sub-Saharan Africa: A case study of a metalworking cluster in Nairobi. *Small Business Economics*, 36(3), 323–335.
South China Morning Post. (May 2002). Ghosts in the machines. Available at: www.scmp.com/article/379463/ghosts-machines
Stein, C. (November 2013). Inside Ghana's electronic wasteland. *Aljazeera*. Available at: www.aljazeera.com/indepth/features/2013/10/inside-ghana-electronic-wasteland-2013103012852580288.html
StEP (Solving the E-waste Problem). (June 2014). One global definition of e-waste. Solving the E-waste Problem White Paper. Available at: www.step-initiative.org/files/_documents/whitepapers/StEP_WP_One%20Global%20Definition%20of%20E-waste_20140603_amended.pdf
(2016a). World e-waste map reveals national volumes, international flows. Available at: www.step-initiative.org/id-2013-12-15-world-e-waste-map-reveals-national-volumes-international-flows.html
(2016b). Israel: Overview of e-waste related information. Available at: www.step-initiative.org/Overview_Israel.html
Sthiannopkao, S., & Wong, M. H. (2013). Handling e-waste in developed and developing countries: Initiatives, practices, and consequences. *Science of the Total Environment*, 463, 1147–1153.
Stubbings, W. A., Nguyen, L. V., Romanak, K., Jantunen, L., Melymuk, L., Arrandale, V., . . . & Venier, M. (2019). Flame retardants and plasticizers in a Canadian waste electrical and electronic equipment (WEEE) dismantling facility. *Science of the Total Environment*, 675, 594–603.
Sullivan, H. (2014). Trash or treasure: Global trade and accumulation of e-waste in Lagos, *Nigeria. Africa Today*, 61(1), 89–112.
Surkes, S. (November 2016). Israel raids Palestinian charcoal plants over pollution. *Times of Israel*. Available at: www.timesofisrael.com/israel-raids-palestinian-charcoal-plants-over-pollution/
Tang, W., Cheng, J., Zhao, W., & Wang, W. (2015). Mercury levels and estimated total daily intakes for children and adults from an electronic

waste recycling area in Taizhou, China: Key role of rice and fish consumption. *Journal of Environmental Sciences*, 34, 107–115.

Tang, X., Shen, C., Shi, D., Cheema, S. A., Khan, M. I., Zhang, C., & Chen, Y. (2010). Heavy metal and persistent organic compound contamination in soil from Wenling: An emerging e-waste recycling city in Taizhou area, China. *Journal of Hazardous Materials*, 173(1–3), 653–660.

Tawil-Souri, H. (2009). New Palestinian centers: An ethnography of the checkpoint economy. *International Journal of Cultural Studies*, 12(3), 217–235.

Tirado-Soto, M. M., & Zamberlan, F. L. (2013). Networks of recyclable material waste-picker's cooperatives: An alternative for the solid waste management in the city of Rio de Janeiro. *Waste Management*, 33(4), 1004–1012.

Tong, X., & Wang, J. (2004). Transnational flows of e-waste and spatial patterns of recycling in China. *Eurasian Geography and Economics*, 45 (8), 608–621.

Tsydenova, O., & Bengtsson, M. (2011). Chemical hazards associated with treatment of waste electrical and electronic equipment. *Waste Management*, 31(1), 45–58.

Tuck, E., & Yang, K. W. (2014). R-words: Refusing research. In D. Paris & M. T. Winn (Eds.), *Humanizing research: Decolonizing qualitative inquiry with youth and communities*. Thousand Oakes, CA: Sage Publications.

UNEP (United Nations Environment Programme). (1989). Basel Convention on the control of transboundary movements of hazardous wastes and their disposal, United Nations Environment Programme/Secretariat of the Basel Convention. Available at: www.basel.int/text/documents.html

UN OCHA (United Nations Office for the Coordination of Humanitarian Affairs). (July, 2011). Barrier update. Seven years after the advisory opinion of the barrier: The impact of the barrier in the Jerusalem area. Available at: https://web.archive.org/web/20160912193655/www.ochaopt.org/documents/ocha_opt_barrier_update_july_2011_english.pdf

US NEWS (August 2014). E-waste in developing countries endangers environment, locals. Available at: www.usnews.com/news/articles/2014/08/01/e-waste-in-developing-countries-endangers-environment-locals

van Leeuwen, F. R., Feeley, M., Schrenk, D., Larsen, J. C., Farland, W., & Younes, M. (2000). Dioxins: WHO's tolerable daily intake (TDI) revisited. *Chemosphere*, 40(9–11), 1095–1101.

Velis, C. (2017). Waste pickers in Global South: Informal recycling sector in a circular economy era. *Waste Management & Research*, 35(4), 329–331.

Velis, C. A., Wilson, D. C., Rocca, O., Smith, S. R., Mavropoulos, A., & Cheeseman, C. R. (2012). An analytical framework and tool ('InteRa') for integrating the informal recycling sector in waste and resource management systems in developing countries. *Waste Management & Research*, 30(9), 43–66.

Wang, C., Gao, W., Liang, Y., Wang, Y., & Jiang, G. (2018). Concentrations and congener profiles of chlorinated paraffins in domestic polymeric products in China. *Environmental Pollution*, 238, 326–335.

Wang, F., Huisman, L., Meskers, C. E., Schluep, M., Stevels, A., & Hagelüken, C. (2012). The Best-of-2-Worlds philosophy: Developing local dismantling and global infrastructure network for sustainable e-waste treatment in emerging economies. *Waste Management*, 32(11), 2134–2146.

Wang, J., Chen, S., Tian, M., Zheng, X., Gonzales, L., Ohura, T., … & Simonich, S. L. M. (2012). Inhalation cancer risk associated with exposure to complex polycyclic aromatic hydrocarbon mixtures in an electronic waste and urban area in South China. *Environmental Science & Technology*, 46(17), 9745–9752.

Wang, J., Liu, L., Wang, J., Pan, B., Fu, X., Zhang, G., … & Lin, K. (2015). Distribution of metals and brominated flame retardants (BFRs) in sediments, soils and plants from an informal e-waste dismantling site, South China. *Environmental Science and Pollution Research*, 22(2), 1020–1033.

Wang, K., Qian, J., & Liu, L. (2020). Understanding environmental pollutions of informal e-Waste clustering in global South via multi-scalar regulatory frameworks: A case study of Guiyu Town, China. *International Journal of Environmental Research and Public Health*, 17(8), 2802–2820.

Wang, K., Qian, J., & He, S. (2021). Global destruction networks and hybrid e-waste economies: Practices and embeddedness in Guiyu, China. *Environment and Planning A: Economy and Space*, 54(3), 533–553.

Wang, Y., He, J., Wang, S., Luo, C., Yin, H., & Zhang, G. (2017). Characterisation and risk assessment of polycyclic aromatic hydrocarbons (PAHs) in soils and plants around e-waste dismantling sites in southern China. *Environmental Science and Pollution Research*, 24(28), 22173–22182.

Wang, Y., Tian, Z., Zhu, H., Cheng, Z., Kang, M., Luo, C., … & Zhang, G. (2012). Polycyclic aromatic hydrocarbons (PAHs) in soils and vegetation near an e-waste recycling site in South China: Concentration, distribution, source, and risk assessment. *Science of the Total Environment*, 439, 187–193.

Watson, I. (May 2013). China: The electronic wastebasket of the world. *CNN*. Available at: www.cnn.com/2013/05/30/world/asia/china-electronic-waste-e-waste/index.html?no-st=9999999999

Widmer, R., Oswald-Krapf, H., Sinha-Khetriwal, D., Schnellmann, M., & Böni, H. (2005). Global perspectives on e-waste. *Environmental Impact Assessment Review*, 25(5), 436–458.

Wigblad, R. (1995). Community turnarounds in declining company towns: A restructuring model. *The Journal of Socio-Economics*, 24(3), 463–475.

Williams, E., Kahhat, R., Allenby, B., Kavazanjian, E., Kim, J., & Xu, M. (2008). Environmental, social, and economic implications of global reuse and recycling of personal computers. *Environmental Science & Technology*, 42(17), 6446–6454.

Williams, E., Kahhat, R., Bengtsson, M., Hayashi, S., Hotta, Y., & Totoki, Y. (2013). Linking informal and formal electronics recycling via an interface organization. *Challenges*, 4(2), 136–153.

Wilson, D. C., Velis, C., & Cheeseman, C. (2006). Role of informal sector recycling in waste management in developing countries. *Habitat International*, 30(4), 797–808.

Wong, M. H., Wu, S. C., Deng, W. J., Yu, X. Z., Luo, Q., Leung, A. O. W., … & Wong, A. S. (2007). Export of toxic chemicals: A review of the case of uncontrolled electronic-waste recycling. *Environmental Pollution*, 149(2), 131–140.

Wu, Q., Leung, J. Y., Du, Y., Kong, D., Shi, Y., Wang, Y., & Xiao, T. (2019). Trace metals in e-waste lead to serious health risk through consumption of rice growing near an abandoned e-waste recycling site: Comparisons with PBDEs and AHFRs. *Environmental Pollution*, 247, 46–54.

Wu, Q., Leung, J. Y., Geng, X., Chen, S., Huang, X., Li, H., … & Lu, Y. (2015). Heavy metal contamination of soil and water in the vicinity of an abandoned e-waste recycling site: Implications for dissemination of heavy metals. *Science of the Total Environment*, 506, 217–225.

Wu, W., Wu, Y., Wu, J., Liu, X., Chen, X., Cai, X., & Yu, S. (2018). Regional risk assessment of trace elements in farmland soils associated with improper e-waste recycling activities in Southern China. *Journal of Geochemical Exploration*, 192, 112–119.

Xinhua Net. (September 2005). E-waste from developed countries flooding towards Guiyu. Available at: http://news.xinhuanet.com/fortune/2005–09/03/content_3437772.htm

Yang, J., Lu, B., & Xu, C. (2008). WEEE flow and mitigating measures in China. *Waste Management*, 28(9), 1589–1597.

Yang, S., Gu, S., He, M., Tang, X., Ma, L. Q., Xu, J., & Liu, X. (2020). Policy adjustment impacts Cd, Cu, Ni, Pb and Zn contamination in soils around e-waste area: Concentrations, sources and health risks. *Science of the Total Environment*, 741, 1–9.

Yao, Z., Li, J., Xie, H., & Yu, C. (2012). Review on remediation technologies of soil contaminated by heavy metals. *Procedia Environmental Sciences*, 16, 722–729.

Yeboah, I. E. (2000). Structural adjustment and emerging urban form in Accra, *Ghana. Africa Today*, 47(2), 61–89.

Yeung, H. W. C. (2016). *Strategic coupling: East Asian industrial transformation in the new global economy*. Ithaca, NY: Cornell University Press.

Yu, J., Williams, E., Ju, M., & Shao, C. (2010) Managing e-waste in China: Policies, pilot projects, and alternative approaches. *Resources, Conservation and Recycling*, 54(11), 991–999.

Yu, X. Z., Gao, Y., Wu, S. C., Zhang, H. B., Cheung, K. C., & Wong, M. H. (2006). Distribution of polycyclic aromatic hydrocarbons in soils at Guiyu area of China, affected by recycling of electronic waste using primitive technologies. *Chemosphere*, 65(9), 1500–1509.

Yu, Y., Zhu, X., Li, L., Lin, B., Xiang, M., Zhang, X., . . . & Wan, Y. (2019). Health implication of heavy metals exposure via multiple pathways for residents living near a former e-waste recycling area in China: A comparative study. *Ecotoxicology and Environmental Safety*, 169, 178–184.

Zahara, A. (March 2016). Refusal as research method in Discard Studies. Discard Studies Blog. Available at: https://discardstudies.com/2016/03/21/refusal-as-research-method-in-discard-studies/#:~:text=Ethnographic%20refusal%20is%20a%20practice,issues%20on%20their%20own%20terms

Zhang, K., Schnoor, J. L., & Zeng, E. Y. (2012). E-waste recycling: Where does it go from here? *Environmental Science & Technology*, 46(20), 10861–10867.

Zhang, Q., Ye, J., Chen, J., Xu, H., Wang, C., & Zhao, M. (2014). Risk assessment of polychlorinated biphenyls and heavy metals in soils of an abandoned e-waste site in China. *Environmental Pollution*, 185, 258–265.

Zhang, T., Huang, Y. R., Chen, S. J., Liu, A. M., Xu, P. J., Li, N., . . . & Mai, B. X. (2012). PCDD/Fs, PBDD/Fs, and PBDEs in the air of an e-waste recycling area (Taizhou) in China: Current levels, composition profiles, and potential cancer risks. *Journal of Environmental Monitoring*, 14 (12), 3156–3163.

Zhang, W. H., Ying-Xin, W. U., & Simonnot, M. O. (2012). Soil contamination due to e-waste disposal and recycling activities: A review with special focus on China. *Pedosphere*, 22(4), 434–455.

Zhao, G., Wang, Z., Zhou, H., & Zhao, Q. (2009). Burdens of PBBs, PBDEs, and PCBs in tissues of the cancer patients in the e-waste disassembly sites in Zhejiang, *China. Science of the Total Environment*, 407 (17), 4831–4837.

Zhao, G., Xu, Y., Han, G., & Ling, B. (2006). Biotransfer of persistent organic pollutants from a large site in China used for the disassembly of electronic and electrical waste. *Environmental Geochemistry and Health*, 28(4), 341–351.

Zhao, G., Zhou, H., Wang, D., Zha, J., Xu, Y., Rao, K., ... & Wang, Z. (2009). PBBs, PBDEs, and PCBs in foods collected from e-waste disassembly sites and daily intake by local residents. *Science of the total Environment*, 407(8), 2565–2575.

Zhao, W., Ding, L., Gu, X., Luo, J., Liu, Y., Guo, L., ... & Cheng, S. (2015). Levels and ecological risk assessment of metals in soils from a typical e-waste recycling region in southeast China. *Ecotoxicology*, 24(9), 1947–1960.

Zhao, Y. X., Qin, X. F., Li, Y., Liu, P. Y., Tian, M., Yan, S. S., ... & Yang, Y. J. (2009). Diffusion of polybrominated diphenyl ether (PBDE) from an e-waste recycling area to the surrounding regions in Southeast China. *Chemosphere*, 76(11), 1470–1476.

Zheng, G., Xu, X., Li, B., Wu, K., Yekeen, T. A., & Huo, X. (2013). Association between lung function in school children and exposure to three transition metals from an e-waste recycling area. *Journal of Exposure Science & Environmental Epidemiology*, 23(1), 67–72.

Zheng, J., Chen, K. H., Yan, X., Chen, S. J., Hu, G. C., Peng, X. W., ... & Yang, Z. Y. (2013). Heavy metals in food, house dust, and water from an e-waste recycling area in South China and the potential risk to human health. *Ecotoxicology and Environmental Safety*, 96, 205–212.

Zheng, L., Wu, K., Li, Y., Qi, Z., Han, D., Zhang, B., & Huo, X. (2008). Blood lead and cadmium levels and relevant factors among children from an e-waste recycling town in China. *Environmental Research*, 108 (1), 15–20.

Zheng, X., Xu, X., Yekeen, T. A., Zhang, Y., Chen, A., Kim, S. S., ... & Huo, X. (2016). Ambient air heavy metals in PM2. 5 and potential human health risk assessment in an informal electronic-waste recycling site of China. *Aerosol and Air Quality Research*, 16(2), 388–397.

Zuev, D. (2018). Digital afterlife: (Eco) civilizational politics of the site and the sight of e-waste in China. *Anthropology Today*, 34(6), 11–15.

Index

For EU product safety concerns, contact us at Calle de José Abascal, 56–1°,
28003 Madrid, Spain or eugpsr@cambridge.org.

www.ingramcontent.com/pod-product-compliance
Lightning Source LLC
LaVergne TN
LVHW011810200926
848484LV00005B/1132

* 9 7 8 1 0 0 9 4 8 3 6 3 6 *